"As a full-time RVer myself, *The Princess and the RV* by Mary Denmon hit close to home. It's an honest, often emotional glimpse into the beautiful chaos of life on the road-where everything takes longer, repairs never follow a schedule, and plans are just suggestions God chuckles at. Denmon's reflections on breaking free from tradition and embracing the unknown reminded me why we chose this life in the first place. Her raw storytelling made me feel seen, especially the craving for routine in a world that constantly moves. A heartfelt read for anyone living tiny, dreaming big, or just trying to figure out what really matters.

—Ruth Mallon,
Storyteller Promotions

"In life some people learn at an early age, though some never learn that to have friends you must win their "trust" and the Princess definitely knows this as she meets her readers with authentic honesty. People like the Princess that learn how important it is to open one's heart to God's presence experience more of the beauty of life. The Princess gets the tone for each day set first thing by reading devotionals and the Bible making God a part of all that happens to her and her family. However, the Princess learns things happen in God's time not ours. This is a great book for couples considering touring the country in an RV, so they can really learn the intricacies of having an RV. The Princess overcomes her prior concern for trying to live in perfection in her home with the help of her RV experiences. This is a terrific book to see how life can seem like it's falling apart but with family and friends support, prayer and God's help it can all get going in the right direction. It's really a good read.

—Tim Von Dohlen,
attorney, author, Former Texas Legislator,
Pharmacist, and Co-Founder John Paul II Life Center

"*The Princess and the RV* is a heartfelt, funny, and faith-filled memoir that beautifully captures the unpredictable detour of life-and how they can ultimately lead us home. Mary Elisabeth Denmon opens up her life with authenticity, warmth, and humor, taking us from small-town Texas roots to an RV parked somewhere in Pennsylvania. You'll laugh, you'll cry, and you'll be inspired to see that even in life's messiest moments, God is doing His best work. Her journey, both literal and spiritual, is a masterclass in embracing change, releasing control, and trusting in divine guidance."

—JULIE LOKUN, JD

"When your children are little, you can fix things with a hug or a Band-Aid. But when they're grown, and life hurts, all you can do is listen—and pray they find their way through. I talked to my daughter, Mary Elisabeth, almost every day of her RV journey. I heard her excitement at the beginning, and I heard her pain when life took unexpected turns. After all our family had already endured, I hoped God would spare her from more heartache. But He doesn't promise to shield us—He promises to walk with us. And that's exactly what this book shows. Her journey taught both of us. I couldn't give her advice, but I could feel her pain and cheer her on. And yes—having her back home for a season was a blessing. Mary Elisabeth can fix anything... even if she has to crawl under a trailer to do it! This is a beautiful, honest, and inspiring read."

—PAT VON DOHLEN
(Mary Elisabeth's mom)

"*The Princess and the RV* is a heartfelt, honest, and faith-filled journey that had me hooked from the very beginning. While I've only experienced no more than five nights in a row in our own camper, I found the humorous stories completely relatable and delightfully entertaining. The author's willingness to be raw and vulnerable with her emotions made the book incredibly powerful—it's a beautiful reminder that everyone carries unseen struggles.

What stood out most to me was the thread of God's faithfulness woven through every chapter. No matter how hard life gets, this book is a testimony that He is always present, guiding us through. It's both encouraging and uplifting, and I'm so grateful to have read it. *The Princess and the RV* is more than just a memoir—it's a joyful, hope-filled reflection on life, resilience, and grace. An absolute joy to read!"

—Michele Dolezal,
Owner of Hunny's HeARTworks

"*The Princess and the RV* is more than a travel memoir — it's a testimony of love, loss, faith, and resilience. I was struck by the depth of heart and vulnerability Mary Elisabeth poured into these pages. I felt the strength of her marriage to Brandon, the way the RV became both a literal and symbolic journey, and how God's hand wove through even the hardest detours. Her openness about grief and the peace she found in unexpected places is a gift to anyone walking through seasons they never planned for. This story reminds us that while we can't always choose our road, we can choose to keep moving forward in faith — and that's a powerful truth."

—Guillermo "G" Castellanos,
Lead Pastor, NorthRock Church

Don't make the same mistake in life. Love, Your Princess

The Princess and the RV

Cynell & Liam,

Don't miss the pauses in life.

♡, Mary Elizabeth

The Princess and the RV

Finding Peace in Life's Detours

MARY ELISABETH DENMON

XULON PRESS

Xulon Press
555 Winderley Pl, Suite 225
Maitland, FL 32751
407.339.4217
www.xulonpress.com

© 2025 by The Princess and the RV

All rights reserved solely by the author. The author guarantees all contents are original and do not infringe upon the legal rights of any other person or work. No part of this book may be reproduced in any form without the permission of the author.

Due to the changing nature of the Internet, if there are any web addresses, links, or URLs included in this manuscript, these may have been altered and may no longer be accessible. The views and opinions shared in this book belong solely to the author and do not necessarily reflect those of the publisher. The publisher therefore disclaims responsibility for the views or opinions expressed within the work.

Unless otherwise indicated, Scripture quotations taken from the Holy Bible, New Living Translation (NLT). Copyright ©1996, 2004, 2007 by Tyndale House Foundation. Used by permission of Tyndale House Publishers, Inc.

Library of Congress Control Number: 2025909917

Paperback ISBN-13: 979-8-86851-325-1
Hard Cover ISBN-13: 979-8-86851-752-5
Ebook ISBN-13: 979-8-86851-326-8

Dedication

*To my soul mate, my rock, my everything, Brandon.
You get me and for this I am forever grateful.
I love you.*

Table of Contents

1. Small-Town Living 7
2. From Country Mouse to City Mouse. 19
3. Meeting My Prince 23
4. From Rather to Denmon 27
5. Parenting 33
6. The Adventure 47
7. Barbie Dream House 53
8. RV Parks Are Not All Created Equal 59
9. Decisions Ahead 63
10. The Beauty of Humans 69
11. There is No Place Like Home 73
12. Life In a Suitcase 81
13. Time to Evaluate the Situation 93
14. Can We Just Have an Address? 97
15. Oh Christmas Tree 103
16. Back to The Basics 107
17. The Memories We Cherish 111
18. Lessons in Deception 115
19. What Vacation? 119
20. Where In the World Are We? 125
21. Being Honest 129
22. A Permanent Address 133

Epilogue 137

How did we get here? That was a question I asked myself so many times during the days of this unplanned adventure you are about to read in this book. Before we start, I'm going to begin this book with a little insight into myself for those of you who don't know me. (Well, even for those of you who do know me, you probably don't know everything there is to know about Mary Elisabeth Rather Denmon.)

My husband Brandon and I were living in McKinney, Texas, in September of 2021, and we were both anxiously waiting for his company to sell, Meridian Brick; Meridian Brick was born from Boral Bricks and Lone Star Bricks. Brandon was in the business of making and selling bricks for homes and businesses across the US and Canada. Lots of preparation had gone into the final part, the actual sale, of Meridian Brick, and we were both ready for a break.

Finally, the business that Brandon had worked for his entire career had sold, and he was part of the exiting executive team who received a long-term incentive as part of their executive benefits. With this incentive, we had an opportunity to spend some time deciding what was next on his career path, making the choice to not work for the company who bought Meridian Brick. We both needed to get away from this noise of our lives, and the place that gives us that silence is the mountains. Brandon and I wanted to be away from our crowded town of McKinney, Texas, so that we could truly focus on God's whispers and guidance for this next stage of our lives.

Throughout our marriage, Brandon and I have traversed many moves due to him being called to help transform different locations of the brick business. Brandon had been in the building products industry for twenty-six-plus years and with this came many changes, and lots of hours, in our lives. He started at Boral Bricks as a manager trainee and after a year of learning the ropes of how a brick plant runs, he was promoted to assistant plant manager in Henderson, Texas; that was one of the many locations Boral had plants in. This move started the process of what would become many moves in our life, as Brandon transitioned into his next roles moving up the corporate ladder, so to speak.

We had moved from Texas to North Carolina, back to Texas (Henderson, Texas), then to Muskogee, Oklahoma, and then back to Salisbury, North Carolina, and on to Longview, Texas. From there and then to Corona, California, for Brandon to dive into the clay roof tile part of the business. From California, we moved to McKinney, Texas, where we would be for the longest.

> Hearing God's whispers are the key to living out your purpose in life, but they do require action on our end.

We moved about every year to year and a half until we got to McKinney; I can tell you without a doubt that God's whispers are what led us to the peace we felt when we were faced with moving our family for Brandon's newest career challenge. Hearing God's whispers are the key to living out your purpose in life, but they do require action on our end. It would have been easy to stay where we were at each turn in the road, but we both feel life is filled with curves, and the key to fulfilling your purpose is being able to navigate the curves no matter the conditions.

The whispers we were hearing this time in McKinney were different because we both knew this adventure would require a completely new and updated set of navigation tools. We both are creatures of habit and knew that to truly hear God's voice, we needed to be away from our routines.

We soon found a condo in Winter Park, Colorado, and made the necessary arrangements for our dogs and house to be taken care of while we were gone. We had never been to Winter Park before, so this was an opportunity to explore this area. We planned on being gone a little over a week and knew this would be a time of growth and development if we allowed our hearts and minds to truly digest God's guidance.

Once the preparations were taken care of, we loaded our suitcases in Brandon's truck and our GPS for the Rocky Mountains.

When life got crazy, Colorado was Brandon's place to escape and unwind to simply listen in God's playground, so the decision to escape to Colorado came naturally for us both. However, we were not there in Colorado long before a recruiter reached out to Brandon to tell him about a job opportunity in Pennsylvania. We were not ready for that talk and tabled the conversation for about a month until the reality of this idea started to take shape. The recruiter sought Brandon out very intently because Brandon was known in the brick business for his ability to turn the culture of plants into thriving communities of safety, trust, product development genius, and profitability in a short amount of time. This company in Pennsylvania wanted to bring Brandon in as their operations president, but this position would require us to move to Pennsylvania. The opportunity was a big decision to make, one we truly had to pray about and know God's hand was in the details of this decision.

Brandon and I had raised our children, Colton and Logan Elisabeth, and both were out of the house, so we had an opportunity to change our address to the East Coast without disrupting their lives. Raising our children was the most important job we could do in our lives, and I know that the job does not end when the children don't live under your roof anymore. But we had finally hit that time in life most call an empty nest, and we were ready to embrace change, just not the kind of ready that started to take shape at that time in our lives.

In December of 2021, we flew to Philadelphia to see what these insistent people were all about. We landed in Philadelphia, and I knew that I was not in Texas anymore from the weather, but it was fine. The chill in the air required more than a jacket and tennis shoes. Thankfully, we brought our coats and boots to keep us warm because we saw the forecast beforehand that Philadelphia had lows in the teens and highs in the thirties during the daytime. It was Christmas time so at least we could experience some cold weather and fun surroundings while we were in Philadelphia. We lived in North Texas at the time, so we did get colder weather, but nothing like what the East Coast got. Texas can get cold during Christmas time, but you also could get a day in the mid-fifties on Christmas Day.

We stayed at the Ritz Carlton in downtown Philadelphia, and the luxury in this historic building was beautiful. The architectural detail alone was enough to make anyone feel the luxury that is embodied in the Ritz Carlton. The furnishings and the lighting all exude old-school class and demure, while the lobby bar is a must during Christmas. The drinks are served in beautiful crystal glasses, making the ambience soothing and inviting. When you are seated at any location in the bar, you must look up. This is because a beautiful glass dome ceiling surprises your eyes and senses with classic Christmas movies being projected onto the glass. Seeing the classic movie like *The Christmas Story* added an extra layer of the Christmas spirit to our hearts. The champagne vending machine was a nice added touch for those wanting to extend their night, filled entirely with half-bottles of Moet champagne. Truly a fun addition.

We flew home after the meeting, and the reality of this far-fetched opportunity became more of an offer that sounded too good to be true. The people who Brandon met with wooed him with promises of opportunities to lead the decision-making that would transform the brick industry culture. Brandon had built his entire career on creating a culture that thrived, so this sounded like the right fit.

After much negotiation between the company and ourselves, we agreed upon terms, and Brandon was to start his new job on January 31, 2022. Wait, how did that happen so fast? I never thought it would really happen, thinking it was nice to be wanted but we would continue our fun, little hiatus from real life a little while longer in Colorado. You know what they say; it takes a minute for a locomotive to build up enough steam to get going to speed and that once it has built up that steam, you can't just put on the brakes and stop it.

Many things must happen to begin the process of stopping; this is what I felt like with the job opportunity. A locomotive with no clue how to start the whole stopping process, so I decided to just continue putting coal in that engine and get this train on the track. I was no engineer, but I realized rather quickly the train I was on did not come equipped with brakes.

Brandon and I began discussing the reality of what us moving to Pennsylvania would look like and how we would approach this new chapter in our lives. We loosely played around with the idea of buying an RV to allow us the freedom to travel around and see the countryside on the East Coast. We believed the RV would also give us the freedom to research different areas outside of Philadelphia that we might want to live in. The company had a showroom and offices in Philadelphia, but the executive offices were in Redding, about an hour from Philadelphia. I knew from my experience of moving around with Brandon's career that we could live anywhere, and nothing was forever.

I have also learned that we can make all the plans in the world, but God can change them in an instant. This knowledge has helped me through all my life's ups and downs. God has the final say, but I am so glad He does, and I do not.

> I have also learned that we can make all the plans in the world, but God can change them in an instant.

I have been very blessed to have my own interior design business that could travel with me wherever we were located. It was always a bit challenging to find trade pros when we came to a new city/town, but finding a good carpenter, electrician, masonry person, etc., would connect me with others in the construction industry. I also knew that this plan to Philadelphia was, by far, the biggest thing outside of my comfort zone I had ever experienced. However, I trusted my husband and knew that Brandon did not take this sort of decision lightly.

You will also learn from reading this book that I am a planner, and I like to know the plan. I also enjoy an adventure, as this was, but I like to come home to my own bed after the adventure. There is, to me, extreme comfort in being able to decorate for the seasons and having people over for dinner, having a place for our kids to be when they come home. I say all of this to let you know that I am flexible to a point in my plans, but that point ends when I don't have a home to come back to after an adventure, no home base for us to regroup. With this adventure, I was about to learn just how flexible I could be and understand how truly important my home base was.

In reading my book, you will see God knows us more intimately than we can even imagine, and sometimes He must upend our lives to get His message across to us of His plan. God knew He had to completely take away all my comfort for me to be able to listen to what He needed to say. There were many days I just wanted to quit, as you will read in this book, but I did not have that option. My home base was gone, and I had no idea where my new base would be located. We were moving around from RV park to RV park, and the constant change of scenery was exciting at first but then became an exhausting task required to live. I will tell you that this was one of the hardest times in my life, but I did not do it alone. The Holy Spirit was living inside me, giving me the strength and courage to continue moving forward one day at a time.

When we are going through a challenging time in our lives, we don't realize what we are able to overcome until we are on the other side of the track, waving goodbye to the locomotive. Jesus is the ultimate engineer of life, and we have a choice to allow Him or not allow Him to navigate our lives. The choice is one hundred percent up to us. So, what track will you follow?

Chapter 1
Small-Town Living

I grew up in a small town in South Texas called Gonzales and lived there for all eighteen years of my formative-shaping years. What an absolute blessing it was to be from a small town.

Gonzales is a small town in South Texas, sort of between San Antonio and Austin. Most people who are from Gonzales have lived their entire lives there, as well as their parents and grandparents. This should clue you in to the intimate feeling of Gonzales, but also the generational legacy that surrounded most people living there. I happened to be in a unique classification in the town in that my family were first-generation newbies.

Gonzales was made up of about 7,500 people, and if that seems like a small town to you, I can promise you it seemed even smaller to those who grew up there.

Every part of my life was intertwined in who knew me, knew my parents, or knew me because of my parents. If our family needed help taking care of our yard, we would just ask the checkout girl at Boysen's Grocery Store if she knew anyone who could help. She would make a phone call, and we would have someone show up to help in no time. No referral needed because you just simply trusted people, and they trusted you.

I grew up in the Episcopal Church; great church, just not where my friends were going to church. Going to a church where I did not already have built-in friendships forced me to connect with others whom I would not have had a connection with otherwise. I was very blessed to have friends, young and old, who were part of my growing-up years. My friends were people I grew up with who were my age, people who were older than me, my friends by association with others, precious older ladies, and couples in the community.

An impactful, sweet woman who lived in a neighborhood closer to town was Perry Bell. Ms. Bell was a Gonzales icon, and she went to our church and lived in the same neighborhood as one of my dad's work partners. My mom would take my sister and myself to Hidden Oaks neighborhood to a friend of my parents who was a registered nurse, who gave my sister and I allergy shots. We both were allergic to various outdoor influences, such as grass, pollen, dirt, and the like. The sort of thing you cannot avoid.

Looking back on these shots, I think they were filled with saline water because I am still allergic to these things today.

I would volunteer to get my shots first so I could go see Ms. Bell while my mom convinced my sister getting a shot was a good idea. I knew this would take enough time for me to get to visit with Ms. Bell, so I was happy to get my shots and walk two houses over to Ms. Bell's house.

Ms. Bell lived in a beautiful white house, and I can vividly remember how beautiful her furnishings were and how she treated me with respect, allowing me to drink lemonade out of her Waterford crystal glasses. I have imprints of love and joy on my heart, like Ms. Bell's hospitality, that are because of the people I was surrounded by in Gonzales. This community was filled with people who would not hesitate to come and help you, no matter what you needed. The flip side of this environment was that those same people also knew what you were having for dinner before you did.

As a teenager, I thought this lack of privacy was horrible; however, as an adult, I know how very blessed I was to have such a tribe to surround my family with through all of life's challenges and joys. As a child, though, I did not know anything different other than realizing our neighbors faithfully rescued me the time when I was a kid and was covered with fire ants. For those of you unfamiliar with South Texas fire ants, they are literally from the devil and are the most horrific creation known to man.

I am not sure where my mom was when this happened, but I was outside playing on our swing set that just happened to have a beautiful, new dirt pile built right by the slide. I slid down that slide and landed in the biggest pile of fire ants ever. It takes about a half-second for you to realize where the sting on your body is coming from, and in that half-second, those horrible insects can cover your body in complete torture.

While suffering in pain, I ran to our neighbor's back door with tears coming down my face, realizing no words were needed. Our neighbors, Dorothy and Clyde, were the best; they had older children, and my sister and I were sort of like grandchildren for them. Dorothy saw my tears and the welts on my legs and scooped me up to put some kind of magic ointment on my legs. She then gave me a big bowl of chocolate ice cream, and my pain from those horrible insects soon became a distant memory.

Aside from the experience with those vermin creatures, I had a picture-perfect upbringing from an outside perspective. My parents were in love, and they did a good job of teaching my sister and me how to be good humans in the world.

My father, Edward Branch Rather, Jr., was born on July 3, 1943, and man, did God have a firework show for his birth. Daddy was fun, exciting, tough, quick-witted, and a true teddy bear when we were hurt. My dad grew up rather different than most, as his parents, Mary Bond Rather and Edward Branch Rather, lived in Washington D.C. during World War II; this is where my grandfather was stationed and where my dad was born. They lived in D.C. until after the war was over, then decided to move to Hillsboro, Texas, where my grandmother's parents, Bethenia Bond Wilkerson Morrow and Will Morrow (Daddy Will) lived.

Bethenia's husband, Roy Lee, died from the Spanish flu when my grandmother, Mary Bond, was a year old. Bethenia and Will married a few years after Roy's death when Mary Bond was three years old; Will, "Daddy Will," was the only father my grandmother ever knew. My grandparents would go on to have two daughters, Nancy and Bethenia, "Betsy." My grandparents, my dad's parents, were killed in a car accident when my dad was nine years old, and my Aunt Nancy was three and my Aunt Betsy was six weeks old. This tragedy led to my great aunt, my grandfather's sister Mary, being the guardian for all three children. However, Mary was not really cut out to raise three small children. Aunt Mary lived in Washington D.C. at the time of her brother's death, and her life was not conducive to leaving her job in D.C. to come to Hillsboro to raise three small children. Mary never married, as her career was her true love. She was a political socialite who loved the Washington D.C. scene. Her love of the D.C. machine would prove fruitful for her, as she later become President Lyndon B. Johnson's personal secretary.

My great-grandparents were part of the equation that formed the team that raised my dad and his two little sisters, with a few members who were pivotal in shaping my dad's future. The head of the team was Carl, who was their butler and chauffer. At first thought, this sounds presumptuous to think a family employee would raise a child, but Carl was much more than an employee. He was the clay that held the mold together for my dad as he grew up. Carl taught my dad about what it meant to be a man, how to navigate life when you were given a bad hand, and, most importantly, how to build upon a foundation of a family who loved you. It is not lost on me that my dad grew up quite differently than most people I knew, but this also gave him more authenticity in my mind.

My mom, Patricia Carol Wiley, was born on June 11, 1946, and was the middle child of three living a very adventurous upbringing. My grandfather was military and stationed in Santiago, Chile, when my mom was born. My grandparents would later move to the Philippines and then Honolulu, Hawaii, right before my mom was in high school. They were both from the USA, so my mom had family in Texas and Oklahoma to come home to. The best part of seeing family for my mom was going to visit her granny Dora, her dad's mom, in Hillsboro, Texas. I did not miss the detail that both of my parents had roots in Hillsboro, giving life to the saying that God is in the details.

Mom loved Granny and the time she spent with her, as Granny taught Mom to cook, sew buttons back on clothes, and, most importantly, the lipstick rule. Never leave your house without lipstick on your lips; this was her rationalization that glossy lips signaled a put-together woman.

My mom would spend time with her grandmother during the summer in Hillsboro, which formed a special love for Texas for my mom. When it was time to decide where to go to college, my mom did not hesitate when she chose to attend The University of North Texas. The University of North Texas is located in Denton, Texas, and is only a couple of hours away from Hillsboro, so this was an easy decision for her. Personally, I cannot imagine wanting to leave Hawaii to come to school in Texas, but I do understand wanting independence from your parents, which is why my mom chose to go to college back in Texas.

My parents met in college at The University of North Texas, dated throughout college, and got married shortly after they graduated. They began their married life in Hillsboro, Texas, where my mom taught at a nearby elementary school and my dad commuted to Waco to attend law school at Baylor University. My parents lived in Hillsboro throughout my dad's three years in law school. Upon graduating from Baylor Law School, they moved to Austin, Texas, where my dad began his law career as an assistant attorney general, and my mom took a position with the Austin school system teaching elementary-aged children.

After four years of being assistant attorney general of Texas, my parents decided it was time to start a family and wanted to begin this next chapter in a smaller community. Thankfully, a law firm in Gonzales, Texas, reached out to my dad, and my parents, Princess (their dog), and me (my mom was newly pregnant) packed up their life and moved south to start their family. My dad started as a junior lawyer, and the partners quickly saw that he was a firecracker. After two years, Perkins and Dreyer law firm became Perkins, Dreyer, and Rather.

I was born on April 6, 1972, and came home from the hospital in Austin, Texas, to a quaint hotel in Gonzales, Texas, while the house my parents had bought there was being renovated. I am sure I filled our home with lots of love, but God knew I needed a sibling.

My sister, Rebecca Morrow Rather, was born on September 10, 1976. I was four and a half years old when she came home from the hospital, and I literally thought she was my baby. When Rebecca arrived on the scene, my baby dolls were my children, who all had names and personalities. I was the typical only child who created my reality in the world of pretend play. So, when Rebecca was just four days old, I took it upon myself to remove her from her crib and place her in my bed. I mean by this point; I was pretty sure she was my very own live baby doll. Now that I am a mother myself, I imagine that was quite a terrifying realization for my mother to come to when she went to her crib and found Rebecca missing.

Rebecca and I had a truly amazing childhood, living on about ten acres outside of town and with the freedom to explore our backyard without any fear. As I am older now and think back on this time, I realize what a simple life we had to not worry about someone coming to kidnap us or try to hurt us.

Rebecca and I both learned to be each other's best friend, as well as sisters, from the beginning because we lived outside of town, which is where our future friends lived. We each had our own set of friends, but when our mom or our friend's mom could not coordinate a playdate or simply couldn't make the drive to our house or us to their homes, Rebecca and I created our own fun together.

My sister and I explored the woods like it was our own personal Narnia, even using a fort that I'm sure someone else built for our adventures. However, in our minds, we built it, and it became our hideout away from our parents to explore. I am sure our parents were thankful for the woods because not only did they know where we were, but they also got a break from us.

Being the older sibling, I was convinced I was the leader of the games, and that every game I came up with or any idea I had was the best option. Even when Rebecca had another idea, she always deferred to me. I was golden to her, and everything I did or said, she wanted to do and be involved in. I was not always fair to Rebecca when it came to the details of the game, being impatient when she did not understand the rules or wanted to change the rules to suit her needs. But at the end of the day, I would always make sure she was happy and that we ended the day on a happy note.

Once I started middle school, however, I was cursed with the horrible middle school belief that my friends defined my happiness or unhappiness. This led to me getting caught up in pointless middle school drama, which would lead me to be ugly to Rebecca. She never understood why I was mean; after all, she had not done anything wrong to me but just was my safe target to unleash my anger and frustration out on. As an older sibling, I was very type A, so what I said was gospel, but poor Rebecca just wanted to play with her big sister and her big sister's friends.

Looking back on this time now, this was not fair to Rebecca, and I wish I would have learned to direct my emotions in a healthier way. I also wish my parents would have been more insistent in reminding us that Rebecca and I were siblings. We would always have each other while my friends would change and would not always be there.

THE GIFTS YOU DON'T SEE

Now that you have some background on how we arrived in Gonzales, Texas, and our family life, I'd like to share with you about the small-town politics of being a daughter of a well-known attorney in our town and surrounding counties. My sister and I loved going to Dad's law office to play in the library or go to the kitchen and sneak a Coke; this was because Mom did not allow Cokes or any soft drinks or juice at our house, so the Coke was a treat. Dad served as both City Council member and City Attorney for our area at the time. He specialized in oil and gas law, meaning he had more training and was certified to understand the fine details involving oil and gas laws. But when you practice law in a small town, you do a little bit of everything and can apply some practices to your own family.

An example story to give some insight into my dad's character, in general, is one that shaped me for my life to come. I was in eighth grade and had a boyfriend who lived in Florida. Now, let's be honest: A boyfriend who lived in Florida was quite the perfect scenario for my parents, because, well, he was in Florida and not Texas. I later found out that one of my friends was calling my said boyfriend in Florida, but I was curious as to how in the world she was paying for the long-distance bill or how her parents were allowing her to make these long-distance calls to my boyfriend. I was not upset that she was calling my boyfriend, which is surprising, as middle school girls can be selfish and do not always have the best intentions for someone other than themselves. (For any of you that don't understand what I am talking about, at that time, you had to pay for making a call outside of your town; if it was out of state, it was expensive.)

Well, I found out my friend was charging these phone calls to my dad's law office. I am not sure how she came up with this method of payment, but I naively believed this was a logical solution for my problem. This would be the icing on the cake for my understanding of how deceptive girls can be in middle school. We all make mistakes in life, and we either learn from them, or we allow them to define us. I was about to learn from my mistake.

Back in the day, when you made a long-distance call, the operator you spoke to did not even question a twelve-year-old charging a long-distance phone call to a business. Are you kidding me?

Well, I thought if my friend was doing it, I certainly had the right to do it too. After all, it was my dad's law firm, and she did not have a dad who was a lawyer at my dad's law firm. You see where I am going with this, don't you?

As a pre-teen, all we did back then was talk on the phone, which made it impossible for my parents to make, receive, or even place a call from home. This constant phone use led to my parents getting me my own phone line, since my dad needed the main line to stay open so clients could reach him. Well, it was a sweet ride—talking to my love for as long as I desired. However, the phone call ride ended one day after school when my dad's secretary called me in for a meeting. I thought, *Are you kidding me? Just me, no one else, like on my dad's schedule, alone with just he and I?* I knew this was not normal, and I knew in the pit of my stomach why I was being summoned to his office.

I walked in with a smile, acting like I didn't have a care in the world. "Sit down, young lady," my dad said, which wiped the smile right off my face. Now, my dad could have gone many ways with this scenario, but the way he went was the shaping part for me. "I know about the phone calls; I know you did this illegally, and I know you're going to pay back every cent to this law firm." The total for my long-distance love calls was $483.49. Dad said I was grounded until I paid back the money and that Emily, his secretary, would be the banker. So, there I was, grounded the summer before my freshman year of high school, and I had to figure out how to make some serious cash to pay back my dad!

To make money, I washed windows, babysat, cleaned out garages, fed pets, and did whatever needed to be done, taking my cash money in my Zip-loc bag to Emily every Friday. Emily would count the money (even though I had counted it and had a sticky note inside the bag with the amount) and then give me my new total due. It took me six weeks to pay off my debt, but what I thought was going to be the worst time turned out to be the most precious time. I was forced to spend more time with my family, but, more importantly, my dad. I had no idea that this "punishment" would literally be one of the biggest blessings of my life. This time, I believe, was when I began to realize God is in the details of our lives, even when we think all the chips are stacked against us.

I paid off my debt on October 4, 1986, and on November 21, 1986, my mom got a phone call that would change the trajectory of our lives forever. My dad had suffered a massive heart attack, having what we now know to be a massive subarachnoid aneurysm. My dad was overweight, smoked, and had a lot of stress in his life. He took blood pressure medicine but was not treated for his hereditary disease of ADPKD (autosomal poly cystic kidney disease). We now know much more about this disease, and both my sister and I inherited this disease from him. I have come a long way in how to treat myself with this disease after many years of trial and error with medical teams. I am blessed to say today that I am stable and have an amazing medical team in Rochester, Minnesota, that help me manage my disease, with the help of a few local doctors in Dallas.

My dad died in the arms of one of his dearest friends while at a meeting in New York City, in one of the Twin Towers, on November 21, 1986. This was the day the world really stopped spinning for my family. Before my dad left on this trip, he kissed me goodbye, and the last words he said to me were, "Now, take care of your mom and your sister." I had no idea that those words would ring truer than I could have ever imagined at the very young age of fourteen.

The days that followed were filled with details, tears, and unimaginable circumstances that we had to coordinate in just a few short days. I was processing this tragedy the best a fourteen-year-old girl could process it, but in the back of my mind, I knew my role in life just became one of caretaker for my mom and sister. Thankfully, this inner voice I was learning to hear at this time was the Holy Spirit helping me to learn to trust in God's plan, despite what life looked like on the outside.

My dad's funeral was surreal, and I felt like I was living above the situation and not really "in" the situation. The days, weeks, and months ahead were a blur of getting by for me, but I never imagined the pain and devastation my sister was feeling. Back then, counseling was not really a thing, and I can tell you beyond a shadow of a doubt (in my un-professional opinion) that my mom, my sister, and myself ALL needed to be in counseling. We were all continuing in life as if what happened to my dad was a bump in the road. I don't mean we were not sad and that we didn't cry, but we didn't really deal with our new reality.

My mom seemed to handle the situation with busyness and shopping sprees, while Rebecca and I were sort of left to handle our emotions ourselves. I don't say this to throw my mom under the bus because she did what she had to do to survive, while embracing her new reality as a single parent. However, we needed to talk about our new life; we needed to heal and create our new family dynamic. But, instead, we coped.

Many family members came to stay with us and be the solid ground for my sister and I while my mom was doing the best she could to survive the day. Then one day, my Uncle Drew showed up and told my mom it was time to put on her big-girl panties and do this, meaning take care of us and move forward.

Uncle Drew told my mom she did not have the luxury to have a pity party about my dad's passing. She had two little girls to raise, and they needed her, so she needed to get it together. From that point on, I remember life was different, but life was good. Mom went back to teaching, while Rebecca and I figured life out with the help of lots of people. We continued to attend the Episcopal church, and the community we had in Gonzales was the shaping I needed to form my inner drive to be a part of something bigger than myself. I believe that had we moved away at this point in our lives, all three of us would have inauthentically changed. We need to be surrounded by people who were able and willing to help shape us. Being a part of a church and community family makes doing life worth the effort. When "they" say it takes a village to raise kids, this is truer than I could have ever known.

Chapter 2
From Country Mouse to City Mouse

We lived in Gonzales until I graduated from high school, and once I went to Texas State (it was known as Southwest Texas back then), Mom decided it was time for a new start for both her and Rebecca. Mom moved to Lakeway, Texas, and Rebecca started school at Lake Travis Middle School. Lakeway was not very far from Gonzales yet was close to Austin. One can never really know the exact moment when a person's life starts to spin out of control, but Lakeway was the beginning of a very long, hard, and heartbreaking road for my little sister.

Rebecca was surrounded by kids who had absentee parents and too much money on their hands. We lived in a very affluent area of Lake Travis. My mom, Rebecca, and I (when I was home from college) did not live in the gated neighborhoods, but most of Rebecca's friends did. I would say this was very much an area where kids had keys to their home on their person, because they were not going home to parents being able to open the door.

At first, the behavior of Rebecca was not a "big deal": skipping classes or school entirely and sneaking out at night to go chill on the golf course behind our house. Well, as you can imagine, things only got worse from there. Rebecca started using drugs and substances at the age of fourteen, and my mom and I found her in a "drug house" one night after not knowing where she was for twenty-four hours; this led to a private school in San Marcos, Texas. Great idea, though not the solution.

THE PRINCESS AND THE RV

The days were long, and the years were longer, and Rebecca was living a life that could not be sustained. Rebecca was in and out of rehab and in and out of school for that matter too. Rebecca was struggling with the loss of our dad, but without therapy or someone to talk to about his death, I am not sure she even understood where her rebellion was coming from. I cannot imagine being a single parent, but I can tell you my mom was doing everything she could to keep Rebecca from the negative influences that seemed to follow her footsteps. Rebecca did not graduate from high school and found ways to manipulate the best of people out of their money. Rebecca's charm was intoxicating to most, but the charm did not pay for illegal substances. It was not pretty, cute, or fun.

I was in college, trying to help Mom with my sister, but the responsibility landed on my mom, and the juggling act she had to endure was absolutely exhausting. One weekend, I was home from college (I had transferred to A&M by this point to be part of a university that had a true college experience, at least that is what I believed at the time), and mom said she had a date that weekend. Now, she had dates in the past, but they were all weird, eccentric weirdos, I believed. When this man rang the house doorbell, and we answered the door, I think Rebecca and I knew something was different about this guy. He was not scared of Rebecca's attitude nor was he threatened by my lack of interest. This man earned our respect with his attitude toward Rebecca and I, always respecting us but also not allowing us to disrespect him or our mother.

This attitude taught us that this man not only cared about our mom, but also Rebecca and me. He was a gentle spirit that captured all three of the Rather ladies' hearts. Later, this man would become Rebecca and I's dad, and his name was Obert C. Logan. You might have heard of him referred to as "Little O" in earlier football days, as he played football for the Dallas Cowboys back in the real days of football when Tom Landry coached the Dallas Cowboys. Obert learned many lessons during his time as a professional football player, most importantly the value of faith and family. Obert did not have children, so we truly became his own. He was an amazing man, and he took Rebecca and I under his wing immediately. Obert was not weird about it, but rather very respectful and humble.

Rebecca and I came to know Obert as our dad on June 24, 1995; this was the day Obert married our mom, and we were so happy to finally become a family again. My sister and I were both excited and nervous, but mostly excited to have an admirable man that would be present in our lives. I thought this was the missing piece in all three of our lives, and being a planner, I felt like God was finally putting the pieces back together so that we could be a family again. He was the missing piece, but we were about to learn that nothing is forever.

Chapter 3
Meeting My Prince

Rewind to January 26, 1994, at Midnight Rodeo in Lubbock, Texas. I transferred to Texas Tech from Texas A&M in the summer of 1993 and was excited to have a new beginning where not everyone from my high school was around the corner and knew my story. (Well, they thought they knew my story, but most did not really know our story unless you were let into our trusted circle of people.)

When you live in a small town and something as monumental happens as a prominent attorney dying and leaving behind a wife and two children, people talk. Most people thought they knew about our life, but mostly they would make things up to have something to talk about. I learned early in life that just because someone is an adult, this does not mean they have good intentions toward others. My mom had a very close circle of friends who became our family, and they were our people who knew our story and loved us through it all.

Back to college, my major had been pre-med up until this point in 1993, when I had begun to question the reality of being a doctor. I recognized that being a doctor would demand more hours than I was willing to give. I knew that once I had children, I wanted to be fully present as a mom. In my heart, I knew that a career in medicine was not something I was willing to pursue at the expense of being unavailable to my children. My inner bent has always been about making things beautiful and rearranging furniture to fit a space.

My dollhouse I had growing up became my ultimate showroom for my talents of changing styles and colors. I changed my major at Texas Tech from pre-med to interior design three days before I met a very special someone, and another change occurred.

Back to the night of the 26th, my roommate Casey had just turned twenty-one and was finally legal to go to the "real" bars. We had a good-sized crowd there that night to help Casey celebrate her indoctrination into being "legal." I was on a date that night and not really interested in said-date, but was being polite on the date as a good South Texas debutante should be. My date was nice enough, but when he got up to go get me a beer, a blond-headed cowboy, wearing a black cowboy hat and with a glimpse of blond locks, caught my eye, and I didn't blink after that. This cowboy asked me to dance, to which I said, "My date is getting me a beer." At this, he replied, "It's just one dance."

> The Holy Spirit will guide us in life; we just must be open to hearing the voice.

Well, that beer ended up in my hands; it just was handed to me by a blond named Brandon, not the guy I was on a date with. We danced until the last song, and I gave Brandon my number and went home, knowing beyond a shadow of a doubt that I had just met my husband. This is going to sound super cheesy, but you know the old saying when you meet that person, you just know? Brandon was immediately humble, respectful, kind, and honest at our first meeting. The authenticity he showed me from the start truly stopped me in my tracks. I knew that he was not like anyone I had ever dated and that I had met my match.

Brandon called me the next day, and we went out on a few dates before he invited me to his apartment with some friends of his. The inner voice of the Holy Spirit spoke to me with constant assurance that he was different and important to stay with. The Holy Spirit will guide us in life; we just must be open to hearing the voice.

MEETING MY PRINCE

Our relationship was certainly quick to bloom, and three weeks after we met, he asked me if I wanted to go to see his fraternity brothers at North Texas in Denton, Texas. I sort of was freaking out that he asked me to do this, because there was not an event or a special occasion we were going to attend. He simply wanted me to meet his friends. Well, of course I said yes! We left Friday after our classes and drove south for five hours to Denton, and I met A LOT of his friends. We had a great night and ended up spending the night at one of his friends' dorm rooms.

Um, older me was a hundred percent saying, "Really, you stayed in an all-guys dorm on a bunk bed with not only Brandon, but other guys you had just met?" Yep, I sure did. All I can say is the love chemical they say you have when you first meet someone who sparks all those things inside of you, they were not kidding. Overall, you lose your mind, but in a good way.

We woke up the next morning with no real plans, to which I thought we would just chill with his friends, but no, Brandon had other plans. He asked me if ... no actually, he did not ask me. Brandon told me we were going to meet his mom. *WHAT??? Are you kidding me? We just met, and I didn't plan on this. I don't know what to wear, and oh my goodness, I forgot all my makeup in Lubbock.* Insight into me, I was a hundred percent going to be wearing makeup when I went out at night or if I was doing anything (if it were even a thing back then) social media-worthy.

Now since I had realized I forgot my makeup, I was going to buy all new makeup regardless of meeting his mom or not, because we would have gone out that night. So, either way, I was going to Dillard's, but I did not ever have to buy everything at one time. Brandon got a big, huge eye-opener into me when he saw: 1) The bill; 2) The fact that I whipped out my credit card without hesitation to pay for it all; and 3) that I was calm as a cucumber "re-buying" all that makeup. So, now I had makeup and was completely psyching myself up about meeting Brandon's mom.

Brandon had not really talked about his parents much up to this point. I knew that his parents were divorced and that his dad worked for Hunt Oil and was gone a month and home for a month at a time. This happened to be a time when his dad was out of the country.

I don't even remember what I wore, but I know that I was nervous as heck. However, Brandon acted like it was no big thing that he was taking me to meet his mom. So, in my head, I'm thinking, *Well, I guess this is the beginning of the end, because I guess he takes all the girls to meet his mom.* Now that I know what I know, I learned he did not take all the girls to meet his mom; he just knew that he wanted me to meet his mom, not knowing when we would be back in Dallas next.

I met Brandon's mom Linda during that trip, and it was very nice and not the least bit intimidating to meet her. But just know that I pulled out every single manner and debutante trick in the book about etiquette and manners. So, obviously, she was blown away with me with that effort. We left her house the next day and went to one of Brandon's favorite places to eat and then drove back to Lubbock. A whirlwind of a weekend, but one that was very memorable and sealed our fate as a couple.

Now, we had many bumps in the road along the way in our relationship, but that all led up to the day he asked me to marry him at the exact spot he first asked me to dance, Midnight Rodeo. On July 6, 1996, a little over two years after we first met, I said, "Yes," to his proposal, and I was beyond walking on clouds with excitement, nerves, joy, and thankfulness.

How it happened was Brandon had asked all our friends to meet us at Midnight Rodeo that night, so it turned into a surprise engagement party as well. Brandon took me to the spot where we met and got down on one knee, saying a lot of things I don't even remember. He also opened the ring box as soon as he got down on one knee, and that was all my brain could focus on.

We, of course, called my mom as soon as we got home that night to tell her the exciting news. Mom and her friends had a trip planned to Colorado and just happened to be driving through Lubbock the very next weekend, so I was excited to get to show my mom my ring-she was the woman who had been part of the village that raised me in Gonzales, Texas.

Chapter 4
From Rather to Denmon

Brandon was going to graduate from engineering school in December of 1996, and we knew we would move from Lubbock for his job, but we had no idea where. We just knew we would go where the job was and were both willing to go anywhere in the US, except Houston, Texas. It's so hot and muggy there, you'd think the good Lord parked His sunbeam right on top of it.

Brandon and I went to Las Vegas in October with some friends of ours to celebrate his birthday and to go see all the things to do there. Brandon had gotten fitted for a suit before we left, knowing he had a job interview when he got back from Vegas. I stayed a day longer because I was coming back with our friends, and he had to get back for school and the interview. Our whole group went out to eat at a super fancy restaurant that one of our friends had said he would pay for because he had just won twenty-five thousand dollars at the blackjack table, and we all happily obliged.

After dinner, we all went to see Siegfried and Roy's show and then out to a club afterward. Brandon took me back to our hotel room about 1:30 a.m., and said he was going to play cards for about an hour and then would be back. I was exhausted so I went straight to sleep but woke up about 3:30 a.m., and no Brandon; 3:45 a.m., and no Brandon; 4:00 a.m., and no Brandon. At 4:15, Brandon arrived, and I mean arrived. He was living large off the money he won (it was not much comparatively, because we didn't

THE PRINCESS AND THE RV

have money to gamble away, but it was more than we started with) and the oxygen he was breathing in the casino (casinos pump oxygen through their air vents for people to be able to breathe in, giving them more energy). We both fell fast asleep until the alarm went off at 9:00 a.m. for his 10:50 flight. Yikes! We scrambled and got him out the door and into a cab headed to the airport.

Now, one detail I have not mentioned is that I worked for a cellular phone company and was in sales during this time of our lives. I was a pretty darn good saleswoman and made good money, but the real point of this detail is that working with the cellular phone company, we both had free cell phones. Brandon got one, well, because we put him on the "spousal" plan. Don't freak out; everyone knew we weren't married, but everyone knew we were engaged to get married. This was back in the day when you paid by the minute for your cell phone, except that we paid for zero minutes because we were always conducting "business" in our phone calls. Well, not Brandon, but his plan was dirt cheap, and my manager always figured out how to write off his minutes, so we didn't worry much about the bill.

Brandon called me from the airport and asked me if he left his car keys in the room. "Um, no, you did not, babe." So, I went downstairs to see if maybe someone had turned in his keys to the lost and found. Well, miraculously someone had, and I got his keys, but this left him stranded once he landed in Lubbock. He needed to be picked up from the airport, go pick up his suit, and go home to get dressed and get to his scheduled interview. Well, those initial plans did not go according to plan, because someone had stayed up a little too late and forgot their brain in Vegas.

So, Brandon did what Brandon does best; he improvised and didn't miss a beat worrying about the details. Brandon had a friend pick him up who had already spoken with the apartment complex manager, asking them to let him into his apartment so he got home, showered, shaved, and put on khaki pants, a button-downed polo shirt, and his red wing boots. (If you did not own a pair of red wing boots back in the day, well, you just were not cool.) Brandon then jumped back into his friend's car and went to the interview.

I flew home later and went over to his apartment to give him his keys. We then went to get his car at the airport, and the next day, we found out this company who interviewed him wanted to give him a second interview. Fast forward in time, and Boral Bricks hired Brandon as their first manager trainee, as part of their program they started in his name for those who show tremendous potential for their organization.

Brandon graduated from Texas Tech on December 17, 1996, and we left Lubbock in our rearview mirror and drove to Salisbury, North Carolina, on January 2, 1996. We got stuck in Memphis, Tennessee, due to a snowstorm, but Brandon, my dog Avery, and I were delayed just a day and a half until we got to the Holiday Inn in Salisbury on January 5th. Brandon started work with Boral Bricks on January 6th. What a whirlwind.

We got married on March 8, 1997, in Texas. My mom pretty much planned our wedding with me during a few trips to Luling, Texas, that fall of 1996 and then I did two trips home from North Carolina once we had moved. I will say, knowing myself well, I was not really picky about the details of our wedding, so I let my mom have pretty much free reign. I guess that is a testament to me truly wanting to marry the man and not worry about the wedding.

Brandon and I went to Lake Tahoe to go skiing for our honeymoon and had the most amazing five-star time of our lives. Now, I mention the five-star part because when we came home, we were about to experience the reality of paying all our own bills together and everything that comes along with combining every single detail of two people's lives.

It is not a walk in the park, especially when you're too young and naive to think you need to discuss things, like finances, before you get married. Now, before you go off on your judgment aisle, we did discuss finances and our bills and how we thought we would handle our budget before marriage.

… But the reality was very different from the idea on paper; we figured it out though, sort of.

We were also not attending church regularly at this point in our lives, and I would highly recommend you do the opposite of that in your life. As a new couple starting your life together, you absolutely need people to come alongside you to help guide you in the journey we call life. Having a church family truly does make a difference in how you navigate all of life's turns, as I showed with our small-town community when I was growing up.

After we settled down back home in North Carolina, I started working at a cellular phone company in the Carolinas doing the same thing I did in Lubbock, but the key difference was that I did not know where in the world anything was in the area, and this was way before GPS devices or GPS on your phone. So, it took me a minute to get my client base established.

Once I got my client base going, and Brandon and I were figuring out how to live sharing all with each other, God threw an exciting, little twist into our lives. I was pregnant! It was January 31, 1997, and we were NOT married yet, and I was completely freaking out. We still had five weeks until we were to get married in March, and this was definitely not in my plan. I had undergone a type of preventative infertility treatment due to me having severe endometriosis, and we knew that once I was finished with the course of treatment, I would be likely to get pregnant within the first three months of the conclusion of the treatment, which ended in November 1996.

Well (contrary to my parents' beliefs of our holy purity), we found out I was pregnant on the day after Super Bowl Sunday, January 31, 1997. Yikes! I had about a billion questions for God, my mom, and Brandon, but ultimately, we were excited and ready to be parents. We kept it on the down low about me being pregnant until about six weeks after we got married, due to me being so newly pregnant and, well, the wedding, minor detail. I was super thankful I had a job with amazing health insurance, as we had not thought for me to be on Brandon's health insurance because I had a job, and I planned on working for a while before we had children. Wrench, plans, life.

A few months later, we found out we were having a boy, and Brandon was over the moon excited, while I was, well, a bit harder to warm up to the moon part. I had been living with girls since I was fourteen, lived with them in college, and had girl cousins, and I just thought the world of boys was one in which I would not be part of. Once I accepted the fact that I needed to let go of my pink bows and ballet slippers dreams, I never looked back and just was unbelievably thankful we had a healthy baby boy growing inside of me.

Side note: Women are kind of amazing. We can create human beings, though not alone. We do need a tad of help from our spouses, but in all seriousness, we can grow a whole human being inside of us. Then when these little humans are done cooking, we birth them and are expected to take them home and just know what to do and how to make sure they survive outside of our safe and protected wombs. But we were looking forward to meeting our little one soon.

Chapter 5
Parenting

Now, I will push the double-time button and give you the skinny version of the rest of my story, which leads to the real-life lessons I learned about what it means to be a princess in an RV. We lived in North Carolina until Colton, our son, was two and a half months old, then we moved to Henderson, Texas, for my husband to be the assistant plant manager at the plant in Henderson. We lived in the Holiday Inn in Kilgore, Texas, for four weeks until we closed on our first house.

I made plans to create all sorts of memories in our new home, new town, and new church that we began attending. We met many good people during this short time, but the plan to make the memories in Henderson was about to change. Again, I heard the voice of calm and reason in the midst of our interrupted plans. We lived in Henderson for about sixteen months and then moved to Oklahoma where Brandon was going to be the plant manager in Muskogee, Oklahoma. Time to switch gears again and see where God wanted us now.

As I grew older, I would learn that the Holy Spirit had a voice that calmed me in all of life's storms. We headed north to Oklahoma where we found an almost built home that we were able to finish building with our personal details. While our house was being completed, we lived in an apartment for two months until we closed on our house in Muskogee.

We got to spend a little more time in Oklahoma, which allowed for us to meet friends and create relationships with them and their children. Colton started his first Mother's Day Out program in Muskogee at our church, St. Paul's United Methodist Church, and had a friend named Griffin whose mom, Cheryl, and I were friends. Unfortunately, Griffin bit Colton on the cheek at "school" one day, and Colton learned quickly that his mom's rules and other people's moms' rules were not the same. Colton came home from school one day and bit me on the arm, and I bit him right back. Colton did not bite anyone ever again.

At this point in our lives, you would think I would get the hint that making life plans truly needed to come with the flexibility to erase and rewrite the plan. I still had lessons the Lord needed to teach me; I just was a bit slow to get the full picture. We then got the call to move back to Salisbury, North Carolina, (yes, the same place we started) in June of 2000, so we left Oklahoma in July of 2000 and bought a house in Rockwell (outside of Salisbury), North Carolina. We must have needed to go back to be able to have our next little angel, because in May of 2001, I found out I was expecting a baby, and in July, we found out it was a girl. Hallelujah, we had been blessed beyond measure, a boy and a girl! We hit the kid lottery.

Logan Elisabeth was born in January of 2002, and we lived in Rockwell until October of 2003. I remember that October, because I remember that Logan Elisabeth was wearing her Halloween costume when Brandon and I flew to Texas to look for houses in Longview, Texas. Brandon had been promoted to plant manager of the Henderson, Texas plant, and we did not want to live in the same town as the people who worked for him; a little space from your employees is a good thing.

Now before I tell you about our next adventure, I need to give you some important history and insight into my second dad, Obert C. Logan. I know that I mentioned that my mom married Obert in June of 1995, but I have not discussed the tragic ending of their marriage. Obert battled colon cancer earlier in his life, and we all thought he had beat cancer. The doctors were pretty sure the treatments Obert did for his cancer had worked. Well, pretty sure, is not a guarantee. In April of 2002, my mom called me to tell me that Obert's cancer had returned. I did not want to believe this was a reality. I mean, I had already lost one dad, why would God allow this to happen to our family? This thought process is what helped my brain rationalize the fact that Obert's cancer had returned, but he was going to be fine, right? We never know when we are going through a difficult situation how we are going to navigate the situation, but my faith in God is what strengthened my resolve to navigate this next chapter with my mom.

Being that we lived in North Carolina when Obert got his new diagnosis, the distance between my family and my mom and Obert was a challenge. Brandon and I made the decision for the children and I to go live with my mom, so that I could help my mom navigate life while Obert battled for his life to beat cancer. Brandon flew to Texas from North Carolina on the weekends to visit us and then would fly back home on Sunday to North Carolina. This was an exhausting journey for all of us.

Once we all came to the realization that Obert's time left here on earth was coming to an end, Brandon and I had some big decisions to make. Colton and Obert were bonded in a way that was precious to see for anyone who witnessed their love for one another. Grandad (this was the name Colton gave Obert being that Colton was the oldest grandchild) was Colton's reason for existing when Colton was with Mimi (the name she was given by Colton) and Grandad. Grandad and Colton fed the cows and horses and did all things ranching from the first moment Colton came with me to see my parents at their ranch in Texas when Colton was two months old. Now, remember Obert did not have any children, so having a baby was not something he had experienced as a young man. Well, you never would have known that, because Obert would take Colton with him in his truck, navigating Colton's car seat and changing his diapers like he had been taking care of babies his whole life. Grandad was Colton's everything. Obert taught Colton about being respectful to women, working hard, having fun, and always thanking God for every day and everybody in life.

When Obert came home from the hospital with hospice care, my mom and I knew his time on earth was not going to be much longer. But, in true Obert fashion he was going to finish his life on his terms and not how we thought we were planning for it. The first time Obert held Logan Elisabeth in the hospital after I had her, he looked into her eyes and simply said, "angel eyes." He said she was like a gift from God, and she was an angel. This title Obert gave Logan Elisabeth would prove to be his motivation to live his last few days on earth. Logan Elisabeth was born on January 17 and Obert was going to be there for her first birthday party. He was there for her party along with our family and Obert's brother, Jim's, family. We celebrated Logan Elisabeth and thought we were spending our last hours with Obert, only to realize Obert was not only tougher than we knew, but his heart was bigger than we knew. Obert was not going to leave this world with his best buddy in the world in the same house as him. This was the hard decision Brandon, and I had to make. The decision for Brandon and Colton to fly back to North Carolina and for Logan Elisabeth and me to stay in Texas with Mimi and Grandad. Brandon and I made the right decision, because on January 20, 2003, Obert Carl Logan took his last breath knowing that his best buddy would always remember him alive in his Mimi and Grandad's house.

When I talk about how we must choose to see the bright side of all the dark roads, I'm truly speaking my truth. God knew that the loss of Obert would prove to be a devastating blow, as one would expect. God knew Brandon and I needed to live closer to my mom. Brandon's next opportunity came soon after the loss of Obert. Brandon was asked to come back to Henderson, Texas to be the plant manager. Brandon was not completely in love with the idea of coming back to run the plant in Henderson, Texas, (this plant was a very challenging plant on every level of the playing field in the world of bricks) but Brandon knew we needed to be closer to my mom.

Colton started kindergarten in North Carolina and was there for a short three weeks before we flew home to Texas to live with my mom, Mimi, who was a kindergarten teacher. Mimi taught Colton for the first three months of his school years and even has the bragging rights to say she taught him to read.

Once we closed on our house in North Carolina, Logan Elisabeth and I flew to Longview to await the delivery of our things and moved into 1101 Heather Lane, living there until April of 2005. I was super excited to move into this particular house because the previous owners had left their gigantic wooden playset in the backyard. I had dreamed of having one of these as a kid, but my parents insisted on the old school metal playset that gave my sister and I more scrapes and bruises than I can remember. I even cut a piece of my back open while pushing Rebecca in the dual swinger. I mean I was pushing it from the back, but that was how we were going to make it flip completely over. Good thing my parents were not there, and the mean babysitter put a Band-Aid on my back to help it heal, which I have a scar in that spot to this day.

Both Colton and Logan Elisabeth got the opportunity to play on the swing set of my dreams, but only for a year and a half. Brandon then got the call that Boral Bricks wanted him to come and run their roof tile (US Tile-owned by Boral) plant in Corona, California, living a million miles away from the real world and Texas. I mean, was God really even in California? I know that sounds crazy, but I was really being tested with this move. California felt so far away from Texas and from everything that felt like home. Brandon and I booked our flights to John Wayne Airport in Orange County, California, buckled our seatbelts and took off for an adventure. We went to look at the area, and Brandon and I were really wined and dined by the roofing company, deciding it was a once-in-a-lifetime opportunity for his career and our family to live in Southern California.

We (by we, I do not mean Brandon and I but the packers and movers) packed up our house, our dog Avery, and our eight-year-old and three-year-old children, and away we went. We had some amazing adventures and experienced literally the best weather on the planet in California while living there. I used to call the weather fake, because you never had to look at the forecast; it was always just perfect. It was so perfect that when we first moved into our house, I stored all our attic things from our previous house on the side of our house. Literally on the side of our house. When we moved there, I did not know that the attics were unusable due to the heat build-up from the roofs and the fact that houses were not built with floored space to hold or store anything in the attic.

Thankfully, we had a concrete parking area on one side of our house that had a huge gate connected to our neighbor's fence and our house, meant for securely parking an RV. (Was God trying to tell us something?) We stored all of our holiday decorations and other treasures one stores in their attic on the side of our house until we installed overhead storage above our cars in the garage. The moving company people were the ones who suggested we store our attic boxes this way, and I thought they were crazy. Turns out, the weather is fake because not one drop of rain fell during this time, and all the treasures were safe.

We settled into the Southern California lifestyle pretty quickly, finding a church close to our house that would be the foundation for our life while living there. We met lots of people who became dear friends at this church, my ladies Bible study was located here, and Logan Elisabeth went to the pre-school associated with the church. Life was exciting and filled with fun times. We would go to church on Sunday mornings and go straight to the beach after church, loaded with our little grill and food we got at the Spanish grocery store on the way. We would then spend the day in the sun. The kids loved playing in the ocean (now, remember the Pacific Ocean is FREEZING; think 56 degrees Fahrenheit freezing) while Brandon and I got an opportunity to talk about our week, the upcoming week, and whatever else we had not discussed over the week.

Just in case anyone is worrying about our three-year-old at the ocean: When you live in Southern California, your kids MUST know how to swim. Everyone has a pool, and everyone has backyard BBQs. I enrolled both our kids in swimming lessons in our small town in California, taught by former

Olympic swimmers, and they guaranteed our kids would swim by the end of the session. And it worked; our little, bitty three-year-old Logan could jump in a pool without holding her nose and swim the length of the pool like a champ. *Side note: Kids really, really need to know how to swim.*

Southern California was an amazing place to raise kids, but it would not be a forever destination. Brandon got the call in early 2007 that the brick industry wanted him to come run the Southwest division of brick plants, and they needed him to be based in Dallas. Dallas, that is a large city ... OK, OK, OK, yes, I was living in Southern California, so I needed to get over my fear of how big Dallas was. We (again not Brandon and I, but packers and movers) loaded up the same dog and both our kids, and away we went to Dallas, Texas.

To say our family was thrilled to be in Texas might be an understatement, considering we had never lived close to family before. By close, I mean like we could take our kids to grandparents and even great-grandparents to spend the night so we could have a night alone and pick up the kids the next day. It was a new beginning and experience. Eventually, we settled on McKinney, Texas, in the master-planned community of Stonebridge Ranch. We finished building our house and immediately broke ground on a pool in our house in McKinney, going on an amazing ride in McKinney raising our children there.

Brandon worked his way from regional manager of the southwest to regional manager of the east & west to VP of operations to executive vice president of operations. Now, this is an important detail: All of that moving on up in the world did not come without sacrifice and determination on both of our parts. Brandon traveled a lot. And by a lot, I mean like every single week, and he would be gone anywhere from three days a week to the whole week. Now, remember I was at home raising kids, and thank goodness, I had Brandon to bounce ideas off of and the discipline regimens when he was there, but it was mostly me for the physical implementation of said rules, punishments and disciplines, etc.

My family joined a church in Plano, about twenty-five minutes from our house, that was a satellite campus of a much larger church, Fellowship Church, located in Grapevine, Texas. We had visited the main campus in Grapevine during various holidays when we came to visit Brandon's family, who lived near there, so this was a natural choice for us to plant roots in Plano at Fellowship Church. The church would be our church home for the next seven years.

Living in Southern California was an amazing experience, but I did worry about being so far from my mom and sister. During the time we lived in California, Rebecca and her son, Hunter came to live with my mom. I knew this was a good plan for the time being, but I also knew it would come with stressful duties for my mom due to my sister's flexible way of living life.

My mom enrolled Hunter in a local preschool so that he would have some sense of routine while my mom was teaching school. This also gave my mom peace of mind knowing Hunter was taken care of while she was teaching. Rebecca loved Hunter fiercely, she just was not equipped to handle the reality of the daily grind involved with raising Hunter every single day.

I have always felt like change is inevitable, and how we navigate change is the true testament of how we allow our life to unfold.

When I received a call from Rebecca saying that our mom had a date with a man from Austin, my radar went into high alert. I wanted my mom to find happiness in life again, but I was not ready for her to get married again. You read that right; I was not ready. News flash Mary Elisabeth, this was her life, and you are not in control of it. Both Rebecca and I did not take the date our mom had seriously until we found out our mom was flying to San Diego to accompany this man to an event he was attending. Now, we were on high alert.

I was nervous and not really excited about meeting this man, but I did trust my mom. Brandon did have to talk me off my high horse and snap me into reality, because my attitude was not one, I was going to be proud of. I had a big talk with God about this meeting and felt much better after I voiced my concerns to Him. After my talk, I was given the most amazing sense of peace, to which I then knew it was all going to be OK. I was not sure what OK looked like, but I trusted that God had this and if I had learned anything in life up to this point, it is that I am not in control. Oh, I could try and control it, but I would be wasting energy on pointless thoughts and concerns.

We met Tim Von Dohlen in San Diego, California along with his son Chris and family on a beautiful sunny day in August of 2005. The introductions went well, and we all had a great day, and I felt a sense of peace that surpassed all understanding. After this meeting, Mom and I had an opportunity to go on a scavenger hunt of sorts prepared by Tim that took us to the Hotel del Coronado in Coronado, California. Coronado island is close to San Diego and is also where the United States Navy seal teams train to become a Navy seal. The scavenger hunt was a fun way to include me in the budding romance between my mom and Tim. Tim may have not known me very well at this point, but he sure knew how to capture my heart with his carefully orchestrated plan. I was impressed with Tim's creativity, his genuine care for my mom, and his wise way of interacting with me at the beginning of our relationship.

A few short weeks later Tim asked my mom if she had faith, to which she answered she did. On January 7, 2006, my mom, Patricia Carol Wiley Rather Logan married Timothy Donald Von Dohlen in Austin, Texas, surrounded by family and friends. Mr. and Mrs., Tim Von Dohlen were only beginning their chapter as one, but their history is what would turn their fast romance into a story for the history books. Their combination of faith, love of family, inner bent to help others, and drive to make a legacy for their children and now sixteen grandchildren is truly remarkable. They have written a book entitled, "In Life the Journey Is Everything: From the Dump to The Gym and Beyond." I highly recommend you read their book, your faith in God's plan for your life will be restored in ways you never knew could happen.

God sometimes must take you on a journey for you to understand the reason the journey even started. It took me a few years to try to understand why my mom, Rebecca and I would endure so much loss. The addition of Tim to our family was one filled with many emotions, but the lasting result is one of divine thankfulness. I know with every fiber in my body that Tim is the one to finish the story God is writing in both my mom's life and Tim's life. I also know that the progression Tim made from Tim to dad is the most genuine and natural progression, and for this I am forever thankful. Their journey is nothing short of amazing, and they truly give inspiration to all who know them and even those who do not. God is in the business of miracles, so don't ever doubt, not even for one second, that He is not. My life is a testimony to this.

Now, back to the journey Brandon and I were on. We began serving as greeters at Fellowship Church in Plano and eventually, we became the greeting team leaders. This role was the starting point of so many relationships that turned into relationships with people who we did life with. Some of our dearest friendships started from this amazing place, and we are blessed beyond our dreams to have such an amazing group of people we continue to do life with today. Our children were in classes with other volunteers' children and over time, the children began to volunteer within the church. We were an amazing team, juggling life with kids and each other and all the things that fit in between that.

Now to the point of the book: Brandon and I did all this church activity, but we did nothing without the help of the Holy Spirit. Finding peace in life's detours is not just an extension of the book title. It is the foundation of how I learned to navigate change with children and life and all the details life has. I have always felt like change is inevitable, and how we navigate change is the true testament of how we allow our life to unfold. I have had so many people

PARENTING

ask me how I was able to move so often with small children and all that is involved with the details of life. My answer has always been pretty simple: Life is an adventure, and we get to choose how our adventure unfolds. We can choose the safety of the known, or we can embrace the unknown with a spirit of adventure.

Faith, for me, has always played a part in my ability to balance life and all that is woven into our story. My life has been an amazing adventure so far, but it has not come without its challenges. Brandon and I are Christians, and we raised our kids as Christians, but that does not by any stretch of the matter mean we were perfect or faultless. We made mistakes and had fights and got certain things super wrong. But we always knew God was at the center of our lives, and that fact made everything we traversed doable.

While living in McKinney, we, unfortunately, lost a lot of loved ones in our lives. In April of 2007, before we had even moved into our house, we lost Brandon's grandfather on his mom's side, then six months later, we lost my grandmother on my mom's side. A few short years later, we then lost Brandon's grandfather on his dad's side and shortly after his grandmother on his dad's side; this was all while living in McKinney, Texas. (Side note: My "first" dad's parents died when he was nine years old, so I never had the honor of meeting them.)

In the back of my mind, I was thankful we lived in Texas during the time we lost our grandparents, but something inside of me told me that we were not done yet. I am not sure if that was because I had experienced so much loss in my life before then, and this was my inner bent, or if the Holy Spirit was preparing me for what lay ahead.

One of my biggest rules in life is to always make sure my loved ones know where we stand. I don't ever want a tomorrow to come that I wake up to a phone call telling me someone has passed away and knowing I had unresolved issues with them. I say this because I know too many people who have, for whatever reason, decided to cut their loved ones out of their lives.

Now, I understand there are some very serious and legitimate reasons someone may have to cut contact with a family member due to tragic circumstances, and for the sake of one's mental and physical health, it is the best decision. I am talking about an unresolved issue that has gotten blown out of proportion and because of this, it has resulted in hurtful actions costing each party deep pain and separation from one another. God is a miracle worker, and He is in the miracle business. Please, please, please don't be the one who leaves something unresolved only to wake up one morning to a phone call that makes the choice of mending fences for you. Is what you're mad, hurt, upset, or frustrated about truly worth never being able to say goodbye? It is not just a cliche statement to say, "life is short." Life is a gift we are given, so it is up to you to determine how you want to be remembered when your clock stops ticking.

In 2013, a big doozy of loss hit us like a ton of bricks. My little sister was a precious and sweet child but was troubled by so many demons in her adult life (as you learned earlier in the book). When I got a call that she had lost consciousness on October 19, 2013, my world stopped. Rebecca was living with someone I had never met, and I did not even know the person who called me that night about Rebecca, but I would soon learn more than I ever wanted to know about this caller and the reason he called me.

My mom and I had just had lunch with Rebecca on her birthday, September 10th, and she was living by herself in Dallas and seemed to be happy and reasonably healthy. I did not see Rebecca often, due to different schedules and it was hard to pin Rebecca down to meet, so to have just seen her, and in good spirits and health, was such a gift for both mom and me.

Sadly, Rebecca Morrow Rather passed away from this life into Jesus's arms the day after my precious husband's birthday. Rebecca had a sub arachnoid brain aneurism that was a result of her having ADPKD (the kidney condition we have). We know now these can be detected before they become an issue, but one must be willing to invest in their health to follow the steps of detection. Rebecca was not one who invested the time to being proactive with her health. She remained on life support for a little over a day and a half,

PARENTING

and our mom and now dad, Tim Von Dohlen, her son Hunter, and the rest of my family came to Parkland Hospital to say goodbye to our precious Rebecca.

Hunter Logan Whittle was the sunshine in Rebecca's life, the reason for Rebecca living as long as she did, I can guarantee that. In turn, Hunter was the best gift Rebecca ever gave us, and he continues to be a blessing in our family's life today. Hunter is now twenty-five years old and living in St. Petersburg, Florida, living his dream of working in the luxury car industry while working on finishing his bachelor's degree in entrepreneurial business. I know that Rebecca would be so proud of the young man he is today.

When life throws you curveballs, you have to learn how to catch them so that you're not thrown into a perpetual state of striking out. I'm not saying this is easy; I am just saying it is worth the effort. We can choose to be the victim of our loss, or we can choose to be thankful we were allowed the time we had with our loved ones. We can also look forward to the blessings God has in store for us, because He wants us to soar on wings of eagles. God created us to be His kingdom builders, and it is not meant to be easy, but we can find strength from life's experiences. We were a changed family after her passing, but we gained strength by the grace of God to move forward into a new way of life.

Chapter 6
The Adventure

*N*ow on the why-we-lived-in-an-RV-part of the story. We knew how to find a house that was built on a concrete foundation and make it a home in lots of different locations. We were just not educated on how to make our house a home when the home was built on wheels and hooked to a truck, going many places without the help of GPS. So, here we go……

It was October of 2020, and Brandon and I were in Cabo San Lucas for the first time. Remember, this was during the height of Covid, and everyone was still in the dark and scared of everything. Brandon got a call from his boss that their company had potential buyers, and the selling of the company they had been preparing for was now becoming a reality; it was time to engage. We got prepared to buckle down and get his company sold.

Fast forward to the company actually selling, and Brandon was offered a job via the new owners, but it was not the avenue we wanted to take. So, we both decided to take some time away from work and evaluate our lives a bit.

We were both just living our carefree lives at that point when Brandon got a call from a recruiter who was looking for a VP of operations for a brick company out of Australia, who came to the United States a couple of years ago to grow their business there. Well, we sort of toyed with the idea of moving again for a job, but never thought it was anything until we found out who it was and that they did not just want any VP of operations; they wanted Brandon. They wined and dined us, and oh they needed to. Guess where they were based? Pennsylvania. What? Um, we have paid our dues and done all the moving, and our kids were grown, so it was time to live that empty nest life. (Whatever that is.) I don't even know where Pennsylvania is. Ya'll, when you're from Texas, your knowledge of the East Coast equates to below the Mason-Dixon Line and above it. There is New York, and then a bunch of little states that I must look on the map to see where they are. Now, I am not belittling these states; I just had never had a need to know where Delaware or Connecticut was exactly.

It was Christmas time at that point, as you learned earlier in the book, so we decided to fly up there to Pennsylvania to "take a look." We were looking in areas near Reading, Pennsylvania. A few of the towns we looked at were: West Chester, Chadds Ford, Kennett Square, Collegeville, and Malvern. The area was beautiful, but boy oh boy, it was different, like night and day different from anything I had ever known. For example, the back roads we took to get from town to town felt like a step back in time into the 1800s, due to the structures we would see and the undisturbed topography. The bridges we went over and under did not even seem like they were safe to travel due to the way they were built. The downtown areas of the small towns were not just historic-looking, but they were literally historic in that most of the buildings still stood as original structures. Some had been redone, but most still were authentic to the late 1800s and early 1900s when the town emerged.

That trip was a whirlwind of people and emotions. We flew home and prepared for Christmas and then decided to go skiing after Christmas in Crested Butte, Colorado, sort of putting this whole crazy Pennsylvania thing on the back burner.

We were sitting in the Secret Stash restaurant in town when the offer came in via email to Brandon. We had only been in Colorado for three days, but now the relaxing part of the vacation seemed like a distant memory. Brandon said the offer was good; it needed some work, but that the opportunity sounded like a fun adventure for us both. Well, I thought not much more about that idea; I mean, me living above the Mason-Dixon line, crazy, right?

Brandon worked out the details, and that crazy idea became a reality as our life kicked into overdrive. I flew back and forth between Texas and Pennsylvania, meeting with Brandon and with our amazing realtor Cary. We looked at so many houses and finally decided to build a house when we didn't find what we were looking for. During the house-hunting portion of my time with Cary, we also looked at new build houses and builders in the areas. After much deliberation between Brandon and I, we decided on a builder we really loved and began the process of building a house in Pennsylvania. We knew this would be about an eighteen-month process, so we needed to make a plan as to how we were going to navigate this part of our "adventure."

In the middle of March was when Logan Elisabeth had her spring break at The University of Arkansas. I was trying to process being so far away from our children and thought it would help both Logan Elisabeth and myself if she came to Pennsylvania to see where her parents would be living. Logan Elisabeth and I got on an airplane and flew to Philadelphia so that Brandon and I could show her our new stomping grounds. We had a great time exploring and seeing the countryside, ending our stay in the penthouse suite in downtown Philadelphia. It was the Saturday before Logan Elisabeth had to go back to college, and Brandon had to go to New York for business, so he stayed in Reading, and we drove to the Logan hotel in downtown Philadelphia. Logan Elisabeth, myself, and my sweet baby dog, Savannah Grace (Avery had since passed so this was my new fur lover) stayed at The Logan hotel; we got upgraded to the penthouse suite, which was an absolute cherry on top of the Sunday for us. Y'all, we had a baby grand piano in our suite and a dressing room bigger than our bedroom and bathroom at home. We also had a balcony that wrapped around half of the hotel. Logan Elisabeth and I both wished we were staying for the weekend, but we had to get up early to fly back to Texas. What an amazing night and time we had in that suite; it was glorious!

All these introductory details are important, because our story is about to get wild.

We all flew home and began the preparations of finding a moving company and all the things involved with a cross-country move. Now, remember I have done this before. We had moved from Texas to North Carolina and Texas to California, but that was when I had small kids and all the details involved with kids, schools, doctors for the kids, and the list goes on. This was different: this was just Brandon and me and our three dogs: Tripp, Kona, and Savannah Grace.

Our house went on the market during the craziest time in real estate we had ever seen in Texas; sellers were getting multiple offers on their homes, and people were paying way over asking price. This is a good thing for sellers, but surprisingly it is also stressful, because you have so many offers with people offering you things like vacations, etc., to get your house, which seemed so crazy to us. We listed our house above what our realtor thought, hoping to avoid some of the crazy, which did narrow down our field to serious buyers. We knew our house would sell quickly, but we wanted to be able to stay in the house until the end of May. Brandon and I had seen this kind of crazy real estate market in California, but not in Texas. We did have a leg-up on the market in that I am a designer, and our house had every single square inch decorated and not a paint chip in sight.

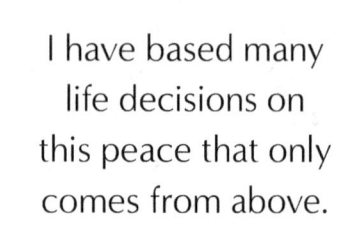

I have based many life decisions on this peace that only comes from above.

We got an offer on the house soon after, but it fell through, leading us to know God had something better in mind for us. Sure enough, our next offer came in way over asking price, and we sold our home of fifteen years in record time. I do have to say that with each move we have done, I have always prayed and asked God to give me peace during the unknown, if the move was the right thing to do or not. I have based many life decisions on this peace that only comes from above.

Thankfully, I did have the most amazing sense of peace about this move in our lives. It really made no sense, except that it was God giving me what I needed to be able to move forward into this next chapter He wanted us in. I can look back on this time now and say that God knew He had to give me this peace, because this move was necessary for us to continue this journey He had for us. Living in the known, comfortable areas of your life does not allow for true growth. True change comes from very difficult decisions and a strong sense of determination to move forward, despite the fear of the unknown that lies ahead.

Chapter 7

Barbie Dream House

*T*his is where our story gets crazy. The day we sold our house and had some money in our pockets, we decided to go look at RVs. I mean what better way to see the countryside we were going to live in, the areas in which we thought we would do life in before living there, right? We started looking small, looking at little bumper-pull RVs, and then the availability of options began to open a big, wide window of possibilities. We learned you could get a washing machine and dryer in an RV; you could even get a second bathroom. This was getting fun, or so I thought.

A little background is that neither one of us had owned an RV or had any experience with RVs, except the one time we rented an RV for a family vacation, and I said I would never do that again unless we owned one. That was a safe statement, right? I mean we would never buy one. The RV we rented was like a small empty apartment with furniture, but no required necessities. We had to bring sheets, pillows, blankets, towels, pots & pans, outdoor chairs, tarps, paper towels, toilet paper, detergent (not that it had a washing machine, but to wash our clothes during our trip) plus all of our clothes and shoes and necessary things. Necessary things that you think you need but learn you don't really need. It took me most of the day to move into this house on wheels, and we had not even left our driveway at home in Texas by 7:00 p.m.

One of the few shining stars of this trip would be that my husband, my children, and myself were all together and no matter what, we would make memories. We pulled out of our driveway in McKinney, Texas at about 9:00 p.m. and made it about two hours down the road toward Colorado. The first night on the road, we stayed in a Wal-Mart parking lot (because we left our house so late in the day and Wal-Mart allows you to park your RV in their parking lot overnight, no questions asked). Brandon and Logan Elisabeth are the hard sleepers of our family, so they fell asleep pretty quickly. Logan Elisabeth slept on the pull-out couch, Brandon and I had the bedroom in the back, and Colton had the bed above where the driver and front passenger sat.

I couldn't sleep, so I thought I would climb up to Colton's bunk to talk, but what we did was people-watch out the window. I was pretty sure Colton, and I watched a drug deal go down outside our window, and I was convinced an ax murderer was going to come into our RV and chop us into tiny pieces, to where we would never be seen again.

Unfortunately, this trip did not start out on a high note, and I was not seeing relaxation in our near future with how we were starting out. RVs are a lot of work, and we were newbies learning along the road, literally. By the end of this family trip, I was one hundred percent convinced we would never buy an RV. However, on May 16, 2022, we did just that, buying a 40-foot fifth-wheel toy hauler. We are from Texas, and we don't do anything small. After we had all the things installed in Brandon's Ford F-250, like a massive piece of machinery used to hitch a trailer to the bed of your truck, we went to pick up our new traveling home.

I decided that this was going to be like the Barbie Dreamhouse I never got from Santa. This was going to be awesome. We hooked that baby up and both about had a heart attack pulling that sucker home on I-35 from Fort Worth, Texas, back home to McKinney, Texas. This trip should normally take about an hour and ten minutes; it took us about two hours. I was completely drenched in sweat when we pulled up to our driveway, and I was not even the one who was pulling/driving this thing. We pulled up and had packers at our house who had already begun to pack our house. I had a designated RV area in our garage with all the things that would go into our RV, having researched everything except the closet part. Have you ever seen an RV closet, not a closet in one of those big diesel-pushers that country music stars ride in from show to show, but a fifth-wheel closet? Well, let me just save you the thought process; it is small, and I don't mean it can't fit your winter coat options small, but it can only fit your whole four seasons of eight articles of clothing and maybe one coat, if it is not a fluffy one.

This was going to be an issue, and I was going to have to get creative. The amount of clothes I had designated for our three-month adventure in the RV, until we found a place where we were going to live in Pennsylvania, was going to have to get paired down, big time, with clothes.

However, I was super organized and ready to tackle this project, like any other project, head on. In my mind, the under-the-bed storage would be for our shoes, because that made sense and would be easy to access. Now I had also researched the mattress that came with the RV. Yeah, that was getting trashed, but not right away; Brandon had another plan for its use. We took that hunk of hard foam off the bed frame and threw it in our front yard for the time being. (We were becoming those people you talk about who had lost their minds with their dreams they come up with.) I brought our awesome new memory foam highly rated mattress in the box into the bedroom, ready to get that baby popped on our bed frame and get rolling with putting our bedroom together.

Surprise, the mattress, all the bedding, and the PILLOWS I had for our new sleeping quarters weighed A LOT! Think having-to-use-your-butt-and-leg-muscles-to-lift-this-under-the-bed-storage heavy. That might be a little issue, but I would worry about it later.

Next task was getting all the cabinets and drawers lined so we could neatly put away all our belongings. Oh, my goodness, talk about having no idea. The space, y'all, is nonexistent. The drawers were small, and I don't mean you-can-kind-of-shove-things-in-them small. I mean if-you-overfill-them-they-are-going-to-break small. My "dreamhouse" in my RV was not shaping up to be as dreamy as I had envisioned. I was determined though; this was going to be an adventure, and we were going to figure out how to live life without the tether of a real house to us. Oh, how God's sense of humor is on point.

Moving is always hectic and no matter how well you plan for it, life happens. This was my fifteenth time to pack up our home or temporary living situation since Brandon and I had begun this cross-country zip code roulette. I had developed a system of organization to help make the process smoother; sometimes it worked, and sometimes we figured out how to make it work along the way. (Now, do not think for one second this system was a computerized program with graphs and spreadsheets. No, this was just my way of organizing our moves that involved a lot of lists.) After fifteen years of living in the same house in Texas, this move would be the biggest one we had undertaken, and we did not take the process lightly or the reality of the details flippantly.

The movers arrived, and I was ready to get them on their mission while I casually loaded our RV. There are so many things that need decisions when you move. Some big, some small, but regardless, decisions must be made. My timetable to get this move done was a lot more generous than the reality I had. Once the house was completely packed up and empty, and the moving trucks pulled away (five eighteen-wheelers to be exact), life got real fast. Talking about moving is one thing, but when you see the house, you raised your children in, made memories in, had birthdays in completely empty, your brain does things you did not put into the equation of moving.

I was trying very hard to compartmentalize the fact that our home was no longer our home, and we had absolutely no idea where our new home would be. I did not have the luxury of truly digesting the fact that we were really doing this; no, I pushed on and continued to pack our house on wheels. Brandon was determined to get on the road the next day, but I knew there was no way that was happening, considering the mountain of things left to be done. This is a time when you use your years of marriage experience and maybe half-truths to prevent the complete unraveling of your husband.

We were scheduled to leave May 27th, and we left May 30th at 1:06 p.m. All three dogs in tow in their car seats and a huge RV attached to our truck headed northeast toward Wilmington, New York. Brandon had signed up for a mountain bike race in Wilmington on June 4th, so we had no time to waste. We made it to Rockwall, Texas, to meet Brandon's dad and wife Marilyn to give them a hug goodbye and do a quick once-over check of all the things after driving our loaded, little house for an hour. (Brandon's parents lived about one hour and fifteen minutes from our former McKinney house, and Rockwall was close to where his dad and wife Marilyn lived.) We had a quick bathroom break, and then our son Colton showed up for a quick hug, secretly, I am sure, to see how we were doing so far on the trip. Seeing your parents' step this far outside their comfort zone is a tad alarming, but Colton had a good game face though, and we left him with promises of seeing him soon.

The day was so windy, which made pulling an RV extra challenging. We had plans to make it to Nashville for the night, but we realized quickly that driving and pulling an RV do not equate to the same timetable of calculating driving distances as with a car. Also, the stress associated with being the passenger in an RV came as a shock to me. I thought I was just going to be able to read my book and play funny videos and games on my phone. The reality was much different. We did not yet know about RV-specific navigation systems, so it was my task to figure out if our route included any non-passable bridges or detours that would not allow us to pass. The other planning task I had was to find a place to park for the night that had the specific requirements for our particular RV. This time was a lot of learning for me and no time to prepare, all part of the wing and a prayer. Meaning, we were completely winging this whole navigation with an RV thing and praying like crazy that the good Lord kept us safe and away from harm. This was our real life, and we had decided to take this life route, very aware that we were not in charge of our route; only God knew the directions.

Chapter 8
RV Parks Are Not All Created Equal

I downloaded all the apps for RV park access in the US, or the ones I could find and got pretty good at calling the RV parks and giving them our basic information. You cannot just show up at an RV park and find any spot you like, drop your stuff, and make your spot home for the night. No, you had to call ahead and make sure they had the accommodations you needed. Not all RV spots are the same size, nor do they all come equipped the same way.

We were a 40 ft. fifth-wheel with three slide outs. (These are parts of rooms inside an RV that you slide into the main body of the RV when you are traveling with the RV, making it seamless; when you park and set up the RV to sleep or stay somewhere, you slide these out, giving you more room in your main rooms). One slide-out was on the passenger side and two were on the driver's side. We also needed a 50-amp electric hookup with full hookups. Now, for those of you who have no idea what I just said, you're normal. I would have never known either had we not had our own traveling house on wheels. Brandon had to, one hundred percent, trust my decisions and judgement on where we would stay, well because he had to drive; this was one of many growing experiences in our marriage. Brandon trusting me with a directional decision was crazy, because I cannot tell you actual directions to save my life. Thank goodness for digital GPS devices.

This information about your RV just gives the parks a heads-up of how big of a spot you need and what accommodations you need when you arrive. It is kind of like booking a hotel room with a king-sized bed, on a high floor, non-smoking (well that should be understood now), and requests for extra pillows, towels, and blankets. Yes, you are correct; that is my standard hotel request when booking a hotel room for my husband and me. We found out quickly that booking these places in advance on our maiden voyage was not going to work, because we were way slower than we thought we were going to be. For example, we might think we would get to Nashville, Tennessee, from Dallas, Texas, in one day, but we were lucky to make it out of Texas in one day.

The newest learning curve we had at the beginning of our journey was the setup when we arrived at the park. It is a whole, big process to plug in all the things (the electricity cord, the water hose hookup to our main water valve to give us water, the black tank drain tube and the grey tank tube hookup to the hole in the ground so we could empty our "liquid" wastes—I know, yuck!) on the RV, making sure they worked. Like, is there enough gas in your propane tanks to use the stovetop? In case you're wondering what powers the RV's heat in the winter, you guessed it: the propane tanks. And when it is cold outside, those tanks are empty in less than a week. We have two of them, and supposedly we have sensors in our main control panel inside our RV that indicate the level of propane. News flash, they don't work. NONE OF THE GAUGES WORK PROPERLY!

Now, when you are married to an engineer, they rule their life on gauges, and when I would tell Brandon the RV gauges didn't work, he would not believe me at face value. "What do you mean they don't work, babe?" How hard is this concept to understand? The gauge said full yesterday, and they say full today, but guess what? The gauge lies because the heater does not work, we have no hot water, and I can't even cook an egg on the stove.

We would both learn how to master the RV park booking process, but we still had a lot of work to do on the figuring out how all the things worked, and what was necessary and what was not necessary when only staying for one night in an RV park. You know so we could get up early the next morning and throw it all in the RV and get rolling down the road. Oh, the lessons life teaches you. The true lesson is if you really learned the lesson the first time, because I can guarantee that you will get another opportunity to see if you're a good student at the University of RV if you don't learn the lesson the first time.

Chapter 9
Decisions Ahead

This is the part of our journey where it got stressful and dicey. I knew we were heading northeast toward Pennsylvania, and I knew we did not know where we would land once we got near the area where Brandon's job was. Unfortunately, I did know that this job was not playing out like Brandon was told it would. The company wanted change, but when change was given, the lead change authority would change what was given, making it completely impossible to implement any procedures, order of process, or accountability for anything done at any level. This was micromanagement at its finest.

Brandon is a loyal, honest, and true person who does not give his word unless he means it. When he agreed to come to work for these people in Pennsylvania, they did not relay the full picture of what their organization operated like. Brandon built his career on hard work and making tough decisions for the greater good of the people of the company he worked for. He has never been one to make anyone feel less than they are because of his position of authority and truly has a gift to be able to understand and empathize with people. People trusted Brandon and counted on him to be a voice for them.

The way this company was operating did not align with Brandon's core principles and beliefs, and the decision to uproot our life was weighing heavily on his mind. Brandon was mentally working through how to navigate the waters in this new company. I do not work in the corporate world,

THE PRINCESS AND THE RV

but I know enough to know that micromanaging anyone is never a recipe for success.

We were in Lancaster, Pennsylvania, at that point, and I didn't have a car, so Brandon got a company vehicle so that I could drive his truck and have some level of freedom when he went to work. For those of you who don't know where Lancaster is or what it is known for, I will educate you. Lancaster, Pennsylvania, is where the Amish and Mennonites have settled and made their home.

I learned that just because the Amish don't have cars does not mean they don't go to Costco and shop. I was very curious about their culture and how they lived amongst society, while choosing to forgo the modern conveniences of living. In time, I learned that families lived on plots of land where they built their homes and worked the soil to grow their own vegetables, fruits, herbs, and more. I also learned, via my curious (ok nosy) nature, that the Amish and Mennonites keep very tidy houses, and their properties look pristine to the naked eye. When I drove by these plots of land, there was not a thing out of place, and the flowers and yards were beautiful.

Now, Mennonites can have cars, and they can also have cell phones. I don't know much else about the Mennonites, except to say they were not nearly as friendly as the Amish people were to me. The Amish people would come to the RV park we were staying in and sell their baked goods and produce on their horse-drawn buggies. The children of the families were also a part of this sales experience and knew just about everything it took to make whatever it is you were buying. It does make one think that if we lived more simply today and held our family closer, doing life with them, what would be different than from our modernized lives now?

Lancaster is where I really started to sink into a sad frame of mind, unfortunately. Brandon was gone during the day, which left me to be with the dogs and walk around the RV park to occupy my time. I had not started back up my design business yet due to many different factors: I did not have any of my industry tools and did not know where to go for supplies; I did not know any contractors that I could trust; and networking to get my business going was not something I saw as realistic, knowing our level of stability in Pennsylvania was low. We were not staying in any place longer than a week or so at this point.

The park was right next door to an amusement park, so that was interesting to watch throughout the day, but even that got mundane to watch. I would take all three dogs to the individual shower rooms so that I could bath all of them. Yes, these were meant for humans, but I just rationalized that I would not be a good human if our dogs were completely stinking it up. Not my brightest ideas, but it was one of many ideas that were unconventional.

One morning, after Brandon left for work, I completely had a meltdown. I missed my family; I missed my kids; I missed my friends, and I was done with this unknown. Brandon knew this job was not going to work, but the reality of the whole situation was overwhelming. I had zero control of the circumstances and zero knowledge of where we were going or what we were going toward. I am a planner, as I mentioned, and I like to know the plan, so this was new territory for me. No plan, no house, no space, nothing about this was making sense. God, why would You have given me such a sense of peace when we were discussing this idea, and then when we followed through with this idea, I was not at complete peace? Had I missed the signs, had I just ignored my real feelings and trusted that this was the right thing to do?

No! I did feel a sense of peace about this move, and I know that we would not have jumped off such a big cliff without the feeling of some sort of security. So, why was I in Lancaster, Pennsylvania, with no one I knew and an RV that was not a home away from home? I called one of my best friends and got honest with her, needing her to help me mentally get over this feeling of despair. I did feel better after we talked, but I was still in a situation that was so foreign and scary to me that I began to get very honest with God. *God, why are we here? God, why did we leave everything we loved and the home we created to get to this point and realize we made the biggest mistake?*

THE PRINCESS AND THE RV

God, I need answers, and I need to feel like myself and not an imposter looking in the mirror, not knowing who is looking back at me. I physically hurt from the angst and sadness I was feeling.

This inner turmoil went on for a few days until I snapped myself out of it and started praying to God every single day to give us answers to where we should go and what we should do. So, we decided to move to a different part of Pennsylvania for Brandon to go see his plants near Pittsburgh. We loaded up all three dogs, folded up our back porch on the RV, and headed to Pittsburgh. We got to Pittsburgh in the evening but by the time we got on the road to the RV place we were going to, things got sketchy.

The road to get to this RV park was narrow, and there were huge trees with limbs that had not been trimmed in what seemed like twenty years, making passing down this road very treacherous. But Brandon did an excellent job of getting us to the park so we both could unhook from the RV and get dinner and a drink after our journey. We were in Pittsburgh for a little over a week and thoroughly enjoyed exploring downtown Pittsburgh. The people in Pittsburgh reminded us more of home, and the places to go and see were more laid-back and just had a good vibe about them. It was fun exploring new places, but I continued to have this voice that reminded me where we were going back to after our day was done. In essence, you can run but you can't hide.

I learned how to cook on a propane-powered Blackstone skillet in Pittsburgh. This was a highlight for me because you could cook whatever you wanted while being outside, so your whole living quarters didn't smell like smoke and crisp bacon. If learning how to cook on a propane-powered griddle outside was my highlight, you can imagine where my frame of mind was. This park also had good laundry facilities, which was nice. This was the first park we had been to that had big washing machines and dryers for washing blankets, dog beds, and lots of towels that needed a big machine to wash everything.

Regarding dog beds, this was the park we almost gave one of our dogs away. Now, before you start accusing us of being soulless dog-haters, let me give you a little background, and then you can make your own decision. I've told you that we were traveling with three dogs. The oldest dog was Tripp, a Malti-tzu who was the sweetest old man but not the sharpest dog around; his sweetness made up for his lack of smarts. The next dog was Kona, a Tibetan spaniel who is sweet when he wants to be, as well as very sneaky and impatient to a fault. The baby dog is Savannah Grace, and yes, you can tell by her name that she is the princess dog who could pretty much do no wrong.

Now, Kona and Savannah together are very smart, and those two together made me want to pull my hair out with their escape antics and food-begging. Kona's kiss of death came when we were in Lancaster, and I did not let them outside quick enough after they ate in the morning. One morning Kona decided he was done waiting on me, so he lifted his leg and relieved himself on Brandon's work backpack filled with all sorts of papers and his computer. I mean he relieved himself on EVERYTHING in Brandon's backpack. Kona decided to pee more than he had ever peed in his life, completely drenching Brandon's backpack, the couch, and the floor. I was so mad, I was calm; you know the kind of calm that you could slice ice with your glare.

After I got the dogs outside and cleaned up Kona's river, I had to decide how to tell Brandon what had happened. Now, by putting them outside, I mean we had a device we would put in the ground with three long leads on them so that the dogs could walk around but not escape. This was the part of the trip where the dogs about had it as well with the RV. I guess we should have realized this once Kona decided to show his true colors to us. Brandon was about as mad as I was when he learned about his backpack, but his anger was not as undetectable as mine was. His anger was more of the decision we should just leave Kona in the park and go about our lives. Of course, we did not do this, but Kona was officially put on notice after this stunt.

We all had a change of demeanor when we left Lancaster and hooked up our RV to head toward Pittsburgh, but I am not sure what that demeanor was; it was a changed one though.

Chapter 10
The Beauty of Humans

Something I studied while moving from RV park to RV park were the people. I wanted to know more about what RV people were doing and whether this way of life was temporary or permanent for them. For me, one of the biggest obstacles to mentally juggle living in an RV was the fact that we had no home to go home to. I would say about ninety-five percent of the people I met were traveling in their RVs for vacation. They planned out their trips a year in advance and had been living the RV experience for many years. They really enjoyed the freedom traveling in an RV gave them and the flexibility they had to go to places they would not normally have gone to without their RV.

In Lancaster, we met a couple who had been coming to the park we were at for seven years every couple of weekends; they were like RV knowledge experts, if there is such a thing. I don't even remember their names, but I do remember their faces and for me, the woman of the couple was like an angel. She asked me to come into their RV and showed me all the things they had done and were going to do, but most importantly, she showed me the "hacks." I learned more from this woman in one night than I could have after watching YouTube videos for a week.

I learned what alien tape was—double-sided, thick, gel-like tape that you can put on the back of anything to stick it to the wall; I learned how to fold towels in such a way that I could stack them on a shelf, because they were rolled into a taco. I learned that you had to have at least three dehumidifiers so that the moisture inside did not take over. Also, I learned that you should use Downy detergent to put down your potty drain because it had a nice smell and had a consistency that helped clean the sides of the black tanks that held said potty business.

This might come as a shock to you at this point in the journey, but we did not really know the first thing about treating our black and gray tanks. Thankfully, we learned quickly how important this is. Our RV A/C system did not originally get routed properly, so there was space (there should not have been this space) in the roof area where the tank lines met at some point (where, I have no idea). When it got over about 85 degrees outside, the smell of rotten vegetables was overpowering; it was awful. Time for me to do some serious research on what to treat your tanks with.

I have now learned that everyone has their own personal way they treat their tanks. It is a balance, because you do need to communicate with your partner about when you are going to empty the tanks. If you have just treated the tanks and then you empty them, you have just wasted your time and stuff you used to treat your tanks. I know you're all dying to know what our special concoction was. My tank treatment solutions were Downy laundry detergent, Dawn dish soap, vinegar, Calgon, and Happy Camper powder, which is made specifically for RV tank treatment. This is not magic, nor is it the recommendation you will get from RV "officials," but it was my concoction that worked. I could go on and on about treating these tanks, but let's just say that it was a disgusting necessity that is part of living in an RV.

Oh, one more thing, you must use special toilet paper when doing your business in an RV. I know; I died too when I found this out. I mean can't a girl just have one luxurious thing in her life that she does not have to sacrifice. Nope, you got to get biodegradable toilet paper that is nice and thin and well, yep, no Charmin extra ply soft cuddly bears will be appearing in an RV commercial for this toilet paper. I know you may find this mundane talking about the topic of "waste," but I am here to tell you it is a big deal when living and traveling in an RV.

Another thing I learned about these tanks is that the sink water and the washing machine water are not any prettier than the other. Do you remember when I told you I was excited that you could have a washing machine in these units? Well, don't be getting too excited, because you are not washing more than one towel or one shirt, a pair of shorts, and underwear; socks might be pushing it. What did this mean? Well, it meant I had to do laundry ALL THE TIME, and I had to empty the gray tank often, like sometimes three times a day, when I was doing laundry. Oh, and to empty said tanks, one must go on the driver's side of the RV and open the door where these controls are housed. Some days, these pulls were easy as pie to pull; other days, they wanted to hold on for dear life before they would give for me to pull them.

Also, if you did not empty the tanks after a load or two of laundry, don't worry. You would know you needed to because the water would leak over the top of the tank and drip down through the bottom of the RV. It was awesome, and this is why I got excited about the one RV park having awesome washing machines and dryers.

I don't even remember the name of the woman who taught me all these tricks for living in an RV; I just know she was sent by God at the exact time for me to learn some things I needed to know and will never forget. We will call her the angel of RV hacks.

Chapter 11
There is No Place Like Home

Back to our next destination, below the Mason-Dixon line was where we were headed in our RV. Brandon had made the decision to leave the company (he had been with the company for about five months at this point), not knowing what that looked like. But he did know he needed to get me back to Texas, and I needed my car. We were not the skilled RV people that would pull a car behind the RV attached to the truck. We loaded up our things in Pittsburgh with a plan to leave the next morning for Texas. However, what we did not know was that a huge storm was on its way, and someone was going to hit a pole, knocking out the very transformer that powered the whole RV park we were staying in that night. We do have solar panels on our RV, and we were so excited about these at the beginning of our adventure, but we soon learned that solar does not mean full power; it just means you get a minimum amount of power at a time.

Brandon did a way better job than I did at getting the whole getting-the-RV-packed-before-morning time down. It started to rain about 8:00 p.m., which in my head meant we were going to be delayed on our departure. I should have known my husband does not give up, and he was going to continue packing in the rain and getting all the things done so we could leave. Oh, and you can bet he had an agenda to follow for where we were going to be on a certain day and how long it would take us to get there. About 2 a.m., (this must have been about the time the transformer was

THE PRINCESS AND THE RV

knocked out from someone hitting an electrical pole) the power went out from the storm, which meant we had to power down most things in our RV to at least have the use of our A/C via our solar panels.

Well, another fun thing that the solar panels would trigger is this stupid beeping sound, letting us know the solar was only going to last for so long. Think of a smoke detector at 2 a.m., as a type of beeping noise. So, I was now thinking for sure we were not going to get out of there early. At this point, you would think I would know my husband, but apparently, he was possessed by some crazy, unrealistic RV demon.

When morning sort of dawned, I told Brandon I would go take a shower in the shower house and get ready for the day. If you know me, I am usually one who gets dressed in hair and makeup each day. However, RV life is very different from real life. I say real life because I felt like I was living in an alternate universe at this point. I was not the only one who had the bright idea to use the shower house early. Every shower was occupied, and it went without being said that a long shower was not a luxury one had that morning from looking at the women waiting to shower after someone was done. I did the best I could in the semi-dark hours of the morning and got dressed, then took the dogs to the potty and then fed them.

We were on track to leave about an hour later than planned, which I thought was pretty good; Brandon, however, did not. In his mind, he needed to make up time. We still had not learned the art of taking breaks when you're pulling such a monster behind you, and we had not learned that you just simply cannot drive as if you are in a car when you're pulling an RV.

We needed to stop for gas in Cincinnati, Ohio, and I had to go to the bathroom, so I hopped out to go to the bathroom while Brandon got fuel. When I was walking out of the next-door gas station (I had to go there because the one Brandon chose didn't have a bathroom), the attendant asked me if I was OK—though everything was fine with me. I said yes, wondering why he asked me that and soon found out. Brandon had jackknifed the RV and busted out the back window of his truck with the neck of the

RV. If you have never done this, it sounds like a gunshot going off. All the dogs were in their car seats in the truck, so this was a complete shock for them. I was the only one not in the truck when this happened, so I was the only one not in shock.

Brandon was in shock by what happened but was still thinking we were going to get to Bowling Green, Kentucky, that night. I had to calmly let Brandon know that we were not going to Bowling Green; we were going to stay in Cincinnati for the night because we had glass in every corner of his truck and didn't have a back window. Thankfully, we found a car wash with vacuums and a place that would replace his back window the following morning. It took us a good hour and a half to vacuum everything, with me knowing we still did not get it all but realizing Brandon needed to park the RV and be done pulling the RV for the day.

We found the headquarters of the Motorcoach Association RV Park in Cincinnati. We had previously joined the Motorcoach Association and just like that, we were able to park in the headquarters park for the night. It was glorious because it was not filled with dirt. Usually, RV parks have a lot of dirt, which is fine except I was so tired of getting my shoes dirty and walking the dogs in dirt and getting them all dirty. So, to see all that blacktop was beautiful to me. Brandon was oblivious to it, as he should have been. All I know is that for me to be able to walk outside the RV and not hit dirt on my first step outside was so awesome.

Brandon had an appointment the next morning to drop his truck off to get the back window fixed, and then we would be off. Once we were back on the road, we both agreed that we weren't on the fast track to getting back to Texas. Brandon and I made it to outside of Little Rock, Arkansas, that day and found a really great place to eat dinner in downtown Little Rock, making life seem a tiny bit normal. As a side note, my aunt and uncle live in Little Rock, and we would have normally called them and stayed with them, had dinner with them, etc. Yet we both decided that we were not in a good frame of mind for anyone to entertain us, so we spared them the uncomfortable scenario.

THE PRINCESS AND THE RV

The next morning, we were headed for the glorious state of Texas. The first RV place we stayed in Texas was in Rockwall, Texas, and it was all blacktop, which was nice! I mean, like the nicest place we had stayed since leaving McKinney, Texas, in our RV. The RV park had a resort-style pool, and the bathroom and shower areas were all marble, with pretty mirrors. Most importantly, they were clean. They had a game room, and it was on the lake. Ok, maybe I was just happy to be in Texas, but, no dirt, which meant walking the dogs was not as crazy. The pool was also super nice and not overly crowded, making it relaxing.

By this point, I had a big life realization that walking three dogs was too much for me. I know it is not mind-blowing information, but at this point, I was glad to be able to mentally digest new attributes about myself that did not involve a terrible scenario happening to learn it.

We spent the 4th of July at this RV park in Rockwall, and Colton came to see us and brought his friend from college as well. What a change. Our son was there with his friend, and we were in an RV with our three dogs; this gave new meaning to close quarters, but it was fine. We were in Texas, our son was with us, and there was no dirt. Brandon and I stayed at the RV park for five nights, and on one of those days, I got my car back. Side note, my car had been transported from McKinney to the car dealership in Pennsylvania. Brandon and I decided the safest place to transport my car was to its dealership, because we were unfamiliar with anything in Pennsylvania. I coordinated getting it back from the dealership in Pennsylvania and had them deliver it to our friend's house in McKinney. Happy day getting my car back.

The next day, I got up super early and got dressed (like put makeup on and fixed my hair) because I was going to get my hair done. It was about two weeks overdue, which if you're a blonde, that is two weeks too long. My hair was amazing after it was done, and all was feeling sort of right in my world. I knew we were going to have to leave our little slice of no-dirt heaven soon, so I savored the win.

Next, we went to Hidden Cove outside of Frisco, Texas, near Lake Lewisville: this was the headquarters of dirt, bugs, spiderwebs, and turkeys. Yes, one afternoon we looked outside our RV, and there were turkeys, plural, on our BBQ area right by our campsite. A mom and her babies, which meant I am sure Dad Turkey was nearby. Why? Why is this my life? All the RVs in the whole park, and they chose our site to hang out. I wanted to take my mind off the crazy, so I decided I would take a shower that was not a RV shower and go to the bath house. Bad decision: There were spiderwebs and a spider the size of Dallas in that bathhouse. Now, I grew up in the country, and I know all about the great outdoors. I just don't want to shower in the great outdoors. Thankfully, I survived, and we moved to the Lewisville area next.

We were moving to different areas in the Dallas-Fort Worth (DFW) area to get a feel for the area before committing to looking at houses to buy. Even though we had lived in McKinney for fifteen years, we still needed to scope out the different areas in the metroplex. We wanted to be somewhere different than McKinney; we just did not know where that was in DFW. We figured out that the campsites near any body of water had a maximum number of days you could stay there, so when we had reached our maximum amount of days, it was time to move, again. (This was one RV spot I would not miss.)

A little sidebar here: I say we moved next to whatever campground. Do y'all realize that when we moved campgrounds, that meant we had to break down our outside house area and inside house area? On the inside, I made sure nothing was in the way of the three slide-outs, so that when we pushed them in (not physically pushed them in, y'all, as there was a button and it did work), nothing got crushed or broken or sucked into the oblivion of wherever all the things go when you push those slides in. I also like to have everything in its place, so this was a learned task.

For tying everything down on the inside, I figured out over time that the dirty clothes hamper cannot just stay where we keep it in the corner, because the dog bed in front of it would squish it to the point of making it not fit in

THE PRINCESS AND THE RV

the corner. I also learned that my shoe holder that fits over the door probably should be taken down when traveling, because if not, I'm going to have a big mess when we do finally stop. The bathroom was a biggie because I had installed these shelves to hold my "beauty care," and these things could literally jump out of the shelf and crash to the floor. The inside of the shower needed to have everything on the floor of the shower, or you might just have bottles of shampoo and body wash all over the shower floor next time you want to take a shower in there.

You understand the point about the inside; now let me tell you about the outside.

The outside had an order in which we liked to load things. Brandon liked to load the Can Am first, because we could get that tied down and then put things around the Can Am. We had a back porch that when you unhook it from its tension rods becomes a ramp to drive the Can Am into the back of the RV. The first time we did this was a complete nightmare, but we got it in there, and we even found the L track hooks to hook the tie-downs into. Y'all, if you do not know what in the world an L track is, don't worry; I did not either and got a whole lesson on the different types of hooks for said track system. These slide-in tie-downs are not all created equal. An L track is a tie-down system that some trailers have with about four tracks that allow you to slide in these hooks to the exact point you need for whatever it is you are tying down.

Side note: We did not know you needed these very important and very specific hooks until about two days before we were going to leave on our maiden voyage. We searched all the hardware stores, the RV dealerships, the motorsports dealerships, Amazon etc., and we did find them, but to get them in two days was the challenging part. We finally found a vendor on Amazon who overnight-shipped them to us. Those stupid, little hooks were like gold when they arrived.

Now that you know about getting item number one into the RV when packing up to hit the road, I am sure you are just on the edge of your seat about the next steps.

Next, we would load up our chairs, hummingbird feeders, tables, grill, dog hook, and, lastly, the rugs. Once we got those rugs (outdoor-specific ones made of waterproof plastic and straw-like material, measuring nine feet by twelve feet), the dirt became manageable—they acted like a mat before the mat when entering the RV. The last things that got loaded were the water hose that attached to the water line, the hose that emptied the tanks, and the electricity. In the summer, electricity is definitely the last thing you unhook (because in that heat, being able to step into a cool RV and lower your body temperature feels like a gift).

Inevitably, you would forget something inside before you got on the road, or you needed to check something to make sure it was done. Like you needed to make sure any sliding doors were locked into place, and the glass shower door was locked, or you were in for a big surprise when you arrived at your destination. (Why they made the shower doors out of actual glass, I will never know.) Last thing was backing your truck into the gooseneck hitch and making it click and then hooking in your lights to your RV to your truck, and away you went. Sounds like a dream come true, huh? Little did we know about what our RVing life still had in store for us.

Chapter 12
Life In a Suitcase

*J*ust like a brick-and-mortar house has issues that you need a repairman to fix, an RV has issues that need to be fixed by a professional as well. Our RV was purchased new, so it had a full warranty, and we had a list of items that needed fixing before we left on our trip to Colorado. Pennsylvania was officially in our rear-view mirror, and we were now going to go on an actual fun adventure in Colorado in our RV. Or so we thought.

We have the most amazing friends, Tanya and Heath, who let us unload the things we needed, while not living in the RV, into their house before we took the RV to the dealership to be fixed. We gave the dealership a month to get our items fixed before we had to pick it up and be on our way to Colorado. And, you guessed it, we were once again on a time crunch because Brandon was signed up for the Leadville 100 mountain bike race. We had about three and a half weeks (we told the RV dealership a month to give us some buffer room) before we needed our RV back to go to Colorado for Brandon's mountain bike race.

If you have learned from reading this book, Brandon is an adrenaline junkie and has always had some sort of extreme sporting event on our calendar. The sporting events sort of got started because I had always been a runner and was training for a half-marathon one year. Brandon does not like to be outdone, so he started running, and we did our first half-marathon together. He beat me time-wise, but it was an event we trained for as a team.

This led to us both training for full marathons which then led to Brandon doing triathlons, which I dipped my toes into for a while but then got hit by a car on my bike, leading to four shoulder surgeries. My days of swimming, biking, and running were done at that point, as it was not worth it to me after that many surgeries. I admit I know I was very lucky to only have sustained a broken collarbone after getting hit by a car, but I just decided the days of my extreme sporting events days were done. However, Brandon's were not, and he continued to do Ironman races and then mountain bike racing.

The hours of training and preparation for races and finisher's medals all were leading up to the ultimate athlete event in North America: the Leadville 100 mountain bike race in Leadville, Colorado. One cannot just sign up for this race; one must qualify for this race or be chosen from a lottery system based on qualifying races throughout the United States that you complete within the designated cut-off times. Brandon completed the Silver Rush (a race that is about 60% of the distance of the 100-mile race in Leadville) race in Leadville in 2021, which earned him a coin to compete in the Leadville 100. The extreme sporting events have always been commonplace in our house, so why would living in an RV make this any different? We were about to find out just how different it would be.

We bounced between our friends Tanya and Heath's house in McKinney and Austin, Texas, which is where my parents live. At the time, our daughter was in Spain for the summer but was due to arrive home at the end of July while our son was still local.

Logan Elisabeth was doing an internship in Valencia, Spain, with a group that counseled children dealing with trauma from their pasts or trauma they were currently living with. This internship was possible through the University of Arkansas and part of Logan Elisabeth's degree program, as she was getting her degree in psychology with a minor in child services. Going to Spain was an amazing opportunity for her and led to her decision to not pursue a career in helping children with trauma. As a side note, having an opportunity to immerse yourself in another culture truly gives one perspective in life and allows you to be able to process what it is you truly want to bring to the world.

We had scheduled a trip to Florida, right before we were to go to Colorado, to see our nephew and his girlfriend. Our daughter's boyfriend had been in Florida working with our nephew all summer, so the trip had many goals involved with it, including one of our daughter's best friends flying into Florida to spend the week with all of us as well. I was so excited about this trip to Florida. A house that was on the water and had plenty of space with a game area, a pool, a hot tub, balconies, and a bathtub. I am a bath girl, and I had not had a bath in over three months. I know that sounds trivial, but a bath is therapy for me where I can completely shut the world out and Zen out in a bath with a glass of wine. Yes, this week was going to be amazing.

We landed in Tampa, Florida, and waited for our daughter's boyfriend to arrive to pick her up. They had not seen each other in over two months, so it was a very exciting time for them. They reunited with joy, and we found our rental car and away we went to Apollo Beach. We arrived at an amazing house and were set for a week of fun in the sun. However, at this point, there was a small part of me that was beginning to wonder where we might land next after this trip. But that was for another time, right?

The week flew by, and we created amazing memories together. Our daughter and her boyfriend headed back to Texas in his truck and Regan, one of our daughter's best friends, Brandon, and I loaded up to head to Tampa International Airport to fly home. As a side note, are you wondering where our dogs were at that point? Our two boy dogs were staying with our pet sitter that truly would become part of our family, while the princess dog was staying with our amazing friend, our friend Tanya's mom, Kathy. Kathy was staying at Tanya's house, who was in Colorado, and graciously offered to keep little Savannah Grace.

The dogs were separated because Savannah Grace is a trained service dog and is trained to detect nerve pain that I have. Savannah has been like

THE PRINCESS AND THE RV

a true family member since we got her and had never stayed with anyone except these friends who knew her and how she was accustomed to living. For example, she sleeps in the bed with you, period, and has free reign of the house because we fully trust her. The boy dogs were not to be trusted, and we were not about to have anyone keep Kona on the off-chance he decided to show his true colors again. Tripp and Kona were best buds, so it was very natural for them to be together and for Savannah Grace to be with our friends.

We arrived back in DFW and prepared to get ready for our trip to Colorado. However, the RV was not ready, which we would soon learn that they never are ready when the dealerships promise you they will be. So, this made the whole timeline for leaving for Colorado not ideal and was really starting to weigh on me. Now, I get that in life there are deadlines, but this whole, "We must leave by this time to make it this far," had we not learned our lesson yet?

Once the RV was done, Brandon went to pick it up, and he hooked it up and set out for home (Tanya and Heath's home) so that we could load it up to leave for Colorado. Unfortunately, a small, little incident happened on his way to load up in the form of a tree limb that caught the awning of the RV. Brandon had a temporary lapse of judgement, forgetting that when pulling an RV, you cannot take the back roads you are used to taking when just driving in your truck. Obstacles that are not a big deal in a truck can become game-changers when pulling an RV. The limb ripped the awning on the RV almost off, which made it necessary for Brandon to pull the truck hauling the RV over, unscrew the awning from the RV, put the awning in the back of the RV, and continue on his merry way.

> We would soon see that when God wants to get your attention, He will.

RV awnings are bolted on high enough that without a ladder, this is a super challenging task to have to disassemble them alone.

In life, God gives you signs along the way, and we had many signs at this point of His plan, but we were not yet in tune enough to see these signs and know what they meant. We would soon see that when God wants to get your attention, He will.

We were back in McKinney at Tanya and Heath's house, who lived a street over from our house that we sold to move to Pennsylvania. We finally began loading the RV in the middle of the afternoon, in August, in Texas, so it was hot as heck, and the cool mountain air could not come quick enough for us. We only had one set of neighbors (who we now know their identities) call the police on us. This was because they did not approve of the sight of our RV on the street, and we were also letting off steam riding around in the Can Am, so they thought they would get us in trouble with this one. Joke was on them though, because if you are actively loading the RV, you have twenty-four hours to load it, and our Can Am (UTV) is street-legal, meaning it was titled, insured, and registered just like any vehicle.

I understand getting upset when people are being annoying/disruptive in a neighborhood setting. I mean we lived in this very neighborhood for fifteen years, because of the safety and HOA rules that kept things in order. So, we were well aware of the rules and truly thought we knew these "neighborly" people who called the police on us.

The "neighborly" thing to do would have been to come talk to us about their concern. It was not secret knowledge that we sold our house and bought an RV and UTV with those who lived around our old street. We lived there long enough to know most people who lived around us, even those around our friend's house too. Anyway, just one of those life lessons that some people are unhappy no matter what you say or do; that is a good tip to keep in mind when traveling through life, especially on those unexpected adventures.

Meanwhile, we "paid" our daughter to drive my car to my parents' house in Austin, Texas, so that it was somewhere out of the way and safe while we were gone to Colorado. Our son Colton picked Logan Elisabeth up in Austin and brought her back to Dallas so that she and her boyfriend could drive back together to Fayetteville, Arkansas, where they were going to college. We left early in the morning on August 6th for Colorado and made it to the KOA in Royal Gorge, Colorado. I am sure it was a beautiful campsite, but we were only there for the night, arriving in the dark and leaving not much after sunrise. We made it then to Salida, Colorado, in the early afternoon, parked, got some lunch, and went to the sock shop.

The sock shop is the original location of Axel socks, and I was tasked with getting the socks for the attire of the ladies in the sherpa group for Brandon and Tanya's husband, Heath. Axel socks are sold all over Colorado and are unique because of their fun colors and sayings on the socks. Most all-adventure towns in Colorado have a few shops that will carry this brand, so to go to the original store meant I had many fun options to choose from.

A sherpa in the extreme sports venue is one who is in communication with the racer and has knowledge of their liquid, nutrition, and health needs along the race course. The sherpa is also made aware of the racer's predicted times and will arrive at certain points along the race course. All of this information allows the sherpa to be located at these points and have the required things the racer might need, as well as anything they may not know they need. By this I mean the sherpa usually intimately knows the racer and has the ability to judge their demeanor by looking at them and having an intuitive feel to how they physically appear at the points you see them at in the race.

Heath had been living in Leadville, Colorado, for about five weeks at this point to get acclimated; this is an important piece to the puzzle for doing this race well. Remember we had been in Florida (i.e., sea level) the week before. So, to say Brandon's lungs were not acclimated by race day in Colorado was an understatement.

I would also say that at this point, I was beginning to get over living in the RV. I know, I know, we just spent an amazing week in Florida with our kids, nephew, and friends, but it was like a trick. Here is a little taste of what it is like to have the security and stability of living in a home and then poof, back to reality in a van down by the river. (This was just a saying my husband and I adopted along the way from a Chris Farley skit on *Saturday Night Live*.)

I was excited to see Tanya and a lot of our other friends who were coming to Colorado to either do this crazy race or cheer on the racers. Most of these people had not seen us since we moved and were in awe of our nomad lifestyle. Some were quick to say it was their dream to live this way, while some were just as quick to say it sounded amazing, but it was not for them.

To those who say it is their dream to live in an RV and travel the country, I have many things to say to you. First, do you still have a home to go home to? Second, how long are you going to plan this trip, because trying to find an RV park on the fly is not as easy as you may think. Third, are you going to rent an RV or buy one? I ask this because it is like moving into an apartment or small home. Remember you have nothing in that RV when you start. Have you thought of things like sheets, towels, pots and pans, pillows, blankets, toilet paper, pantry food and food for the refrigerator, spices, cooking oil, storage containers, utensils, cooking utensils, can openers, mixing bowls, cookie sheets, crock pots? You sort of get the idea. You're moving into this place to LIVE, so just know that you're going to need a lot of things, and then you're going to need a lot of things you would have never thought you needed.

Also, are you OK with knowing that if one person takes a shower in the RV, the next person is not going to get hot water? So, you better plan for those things. Oh, and if you're renting this RV for this amazing adventure, the mattress that comes with any RV is junk; I mean just throw it out the window and research the very best mattress you can afford for the space you can put it in. I would check with the people you rented the RV from before you "throw out the mattress" and leave it with them or offer to let them have the one you bought in place of said junk mattress. If you're buying the RV, same advice. Sleep is important.

Back to our racing story, we went from Salida to Buena Vista, Colorado, where we would stay for about ten days. We had planned to stay in Colorado through the end of August, due to the weather in Texas being so miserably

THE PRINCESS AND THE RV

hot. Unfortunately, we still did not have a plan to find a house, nor did Brandon have a plan as to what he was going to do when he grew up.

Now, I don't mean this sarcastically; I just am literally stating what he would say to people who would ask what our plan was. This was super hard for me, the planner, to relay to others that we did not have a plan and were just figuring it out as we went. I have never been a "wing-it" kind of person, and this experience was not changing this about me. Brandon and I got our space in Buena Vista all lined up, and away he went to go ride parts of the trail before the race. Brandon took our one vehicle with his bike and left the dogs and I in the beauty of the mountains. Don't worry, though; I had our street-legal UTV. Brandon and Heath did this riding a few times a day until a few others they knew arrived and the group grew. They rode their bikes on just about every mile of the racecourse, just not all at once.

My friend Tanya arrived soon after, so I was so happy to get to Leadville and see her and her precious daughter, Claire. I felt like I had not seen my friends in forever, and it just gave me a sense of familiarity to have them close in proximity to me. Race day was fast approaching, and thank goodness Tanya had researched the course, the aid stations, and the roads to get to said stations. This girl, me, had not done one single thing to prepare for the day of the race besides coordinating outfits and making sure we had appropriate snacks and drinks for the day.

Race day dawned, and we arrived in Leadville super early to begin the day. At this current juncture, Brandon and I were good but not great, because this adventure was not panning out to be one of those epic ones. I love this man with every fiber of my being, but that does not mean I was not frustrated with what I felt like was a lack of pursuit of getting us to a stable foundation in our lives. I have always relied on Brandon to be the supporter and the one to lead our family, but I did not feel like he was doing this in the least. I felt like he was figuring out what he wanted to do with the next half of his life, and I was just along for the ride. Though I did not sign up for this ride, I did sign up to love him for better or worse, and this is sometimes what you stand on. I was trying to be a good sport, and I was also trying to not be dramatic before he endured a 100-mile mountain bike race that I thought he had no business doing in the current mind space he was in.

We have not really discussed the part in our story about Brandon mentally going through a mid-life reflection point in his life, not knowing what he wanted to do and where he wanted to do it. To be fair, he was completely blindsided by the company who hired him in Pennsylvania. We were led to believe we had to uproot our life in Texas to be a part of their team in Pennsylvania. A promise from many people for the reality of a string of lies that led us to this point in our lives.

As a man who is in charge and takes charge, Brandon was truly experiencing a roller coaster of emotions. I am used to Brandon being in control, knowing the plan and taking care of the details. Where we are going to live and the details of where we would do life came from Brandon, because I chose to support him and his career at the very beginning of our marriage. This does not mean I did not have any input in these decisions, because I did, and we took life changes as a team. Brandon has always let me be free to do what I wanted to do, with the expectation of making sure certain things that I take care of are taken care of. But we were in a different place now.

The beginning of this race had the normal concerns for the participants and their families, but for Brandon and I this race had the mental weight of where we were in life right then. The day did not go as planned, and Brandon did not finish the race. He made it to mile 67 and was pulled off the course due to time limit cut-offs. For safety reasons, the racers must come to certain mile markers within a specified time to be able to complete the race within the allotted time frame. The race coordinators take the racers' safety very seriously and rightly should, as the conditions of this race are treacherous at the very least. I knew this would be a big mental test for us both, for more than the obvious reasons. Brandon likes to win, and I don't mean just win, but no matter what the situation is, he likes to come out on top. The finish to his race that day would prove to be one of many things he would not win.

THE PRINCESS AND THE RV

We finished our time in Leadville with hearts a little broken, but for different reasons. I was ready to find a home, and he was sad to have not finished a race he had trained so hard for.

We left Buena Vista and went to Ouray, Colorado, a beautiful spot called the Switzerland of America, and this is not an exaggeration. The scenery is breathtaking, and we were lucky enough to get a spot for our RV that backed up to the river, which was truly a peaceful and beautiful spot. A great place to unwind after some mentally challenging days.

Ouray is also where people go who want to explore the mountains in their ATVs or UTVs. We rented a trailer and hauled our Can Am to one of the entrances to these trails; this was the first real adventure we had gone on in the Can Am. The views were breathtaking, but I was pretty sure we would not survive to see the views in reverse, due to the nature of the trails and the proximity to the edge of the mountain. We did make it back to the bottom, but I can one hundred percent say that was one of the scariest few hours of my life. Brandon and I agreed from that moment on that I was not part of the fast and furious club on the side of mountains. Brandon could continue his death-defying adventures while I did my adventuring on more solid ground. Like sipping wine at a quaint mountainside bar.

We stayed in Ouray for almost a week and then headed to Gunnison, Colorado. We were there for almost two weeks and enjoyed our stay with all the farm animals that roamed the campgrounds. Now, this place was not like the place with the turkeys, because these animals pretty much stayed near the owner's house and their pens and stables. They had goats, pigs, chickens, donkeys, and a ram. Quite the comedic group, I must admit.

One morning when I was going to do our laundry, I walked by the goats and thought, *what in the world are we doing? Are we even going in the right direction toward an answer for our lives?* Thankfully, God meets you right where you are, and that morning, He sent an angel to the laundry room for me.

In the laundry room at the campground, I met the sweetest older woman who was on a trip with her husband, and they had not been out in their

RV in three years due to him having cancer. We really connected, and I felt like not only had God sent her to talk to me, but she was able to give me some life advice that helped me navigate my current life situation. I don't remember her name, and I honestly don't even remember where she told me they were from, but that was not important. The gift lay in her willingness to be humble with me and speak words of encouragement and love to me. I needed this more than I knew, and I thank God for her being in that laundry room that day.

We were able to connect with my aunt and uncle who had a house in Crested Butte, Colorado, for dinner one night while at this campground. This was great and a nice treat to see family while we were figuring out our lives. The morning after our dinner, I woke up feeling not so great, and having had COVID before, I knew this is exactly what it was. Having COVID in your own bed in your own home is one thing; having COVID while living in an RV is not ideal on any level of thought. This led to me canceling my scheduled trip back home to Dallas to get my hair done and then to Austin to get my car back from my parents' house.

While recovering from COVID, the realization that my life was in complete disarray was overwhelming. We were both doing our best to keep it together, but as the days moved on, the thoughts got stronger. *When, God, when? Why, God, why?* During this time, my devotion with God was very heartfelt and filled with so many questions and no answers. I felt like the voice of God was so distant, as I craved answers and a plan.

Chapter 13
Time to Evaluate the Situation

Once the sickness had passed and it was time for me to re-book that trip to Austin, I was feeling excited. I had not felt real excitement inside my bones in what felt like months, like Brandon and I both were on survival mode and doing the best we could to find something to be joyful for each day.

You are probably wondering about the whole feeling of excitement for the trip. This is hard to admit, and I don't say these next things lightly. I was excited because I was going to be able to sleep in a real bed; I was going to be able to get dressed in a real bathroom; and I was excited to just be by myself at night. Yes, I was happy to be away from my husband. Yuck, I hate that I said that, but I hate even more that I felt that. However, I promised myself I would be completely honest about this experience for this book, and those were my honest feelings. I did not want to have to talk about trivial things. I did not want my dog, Savannah Grace, to get dirty and bathe her every night to get in bed with me. Yes, she sleeps with us, spoiled rotten this dog is.

I also secretly was celebrating because guess who did not have to do any of the things to load up that RV to get it ready to head back to Texas? Me! And I felt terrible about it but not so terrible that I stayed to help with that. A girl's hair is important! In all honesty, I just wanted a sense of normalcy. However, I knew the moment I got on that plane that my life was far from normal.

THE PRINCESS AND THE RV

My mom came to get me when I landed in Austin from Colorado, and she knew. I am blessed with an amazing mom, and she and I are uniquely tethered due to the life circumstances we navigated together. Pat Von Dohlen looked at me, hugged me, and let me cry. I had discussed details of our life all along the way, but I had not exactly let her know everything. I was faking it and was doing everything in my power to do life and make the best of the situation, still trying to go about doing life like we always had.

News flash, when you do something as drastic as sell your home you raised your kids in, store all of your belongings in storage, buy an RV (because it would be fun to figure out where you wanted to live in this new state that you know nothing about—i.e., this was not a long-term solution), travel the highways to the East Coast, and then back to Texas with the reality that you had no home to return to, life is anything but normal. Mom let me just be in my room at my parents' house, and she and Tim nurtured me the best they knew how. I needed this, and they allowed me the freedom to come undone.

A day later, I gathered my wits, my dog, and my stuff, and my mom drove me to the airport to fly to Dallas to get my hair done. The next plan was to re-group and travel to Arkansas to see our daughter and friends for a few weeks. Brandon was in route from Colorado to Arkansas. I flew from Dallas to Bentonville, Arkansas, and Brandon picked me up at the airport to take me back "home" to our glorious tin home. I had found us a spot at The Creeks Luxury Golf and RV resort in Cave Springs, Arkansas. Sounds fancy, huh? Names can be deceiving, but it was fine for the purpose and only about twenty minutes from the University of Arkansas, where our daughter lived. It was awesome being so close to her, but also heartbreaking for me because I could not answer the question she kept having. The question we got about every three to four days from our daughter was, "Have you found a house?" The answer would become harder and harder to answer, because our daughter looks to her dad and I as the ones who have the plans. Life takes a new turn when the parents have no clue what they are doing but try to act as if we have it all figured out.

TIME TO EVALUATE THE SITUATION

One week into this visit, and we knew that we needed help from an outside source; we needed counseling. Brandon and I both had been led to believe our lives were going to be one way, and we both had been disappointed and were mad at where we were. We only had each other to be mad at, but we were not mad at each other; we were just the convenient participants. Thankfully, Brandon and I found a therapist who could meet with us online and was a good fit for what we needed in taking that step toward healing.

Let me be the first to tell you that deciding to begin marriage counseling is not an admission of failure. It is the best investment you can make on the most sacred decision you make in life. I am not saying it was easy to begin counseling with Brandon. I had a lot of stuff that I had not said and a lot of anger I had suppressed from him. We both had a lot of things we needed to deal with personally and with each other. However, we love each other in a fierce way that runs deep. The kind of love that takes root in your soul and becomes an extension of who you are. This kind of love is worth fighting for but can also be very volatile if left untreated.

We are also both type A personalities, so when we do not continually work through our issues or concerns, our teapots boil over. It took us about three months to come to a point that we were mentally healthy with each other, thanks to the counseling. The difficult part of the equation was the unhealthy view Brandon had toward the company that had completely uprooted our lives in a disingenuous way. Brandon had anger that only he could work through regarding this part of the puzzle, which was hard for me because I was the only one available to deal with the aftermath of the situation.

One evening, Brandon told me that I did not understand the situation because it had nothing to do with me; it only had to do with him. Now, in retrospect, I can see how he thought that, but come on, did he really think this whole thing had nothing to do with me? No, I was not the one the company had lied to. No, I was not the one the company showed their questionable business practices to. But I was the one married to the one who had been dealt that hand of cards. So, yes, it did affect me just as much as it did him, just in a different way.

We both had a lot of work to do to cross over to the other side of the situation and be able to look back and see the lessons and blessings we had been given in this circumstance. Thankfully, Arkansas was the beginning of the healing journey.

Chapter 14
Can We Just Have an Address?

From Arkansas, we went to Flower Mound, Texas, to Twin Coves Campground. This was a nice campground spot, though it did not have any amenities for RVs, such as laundry facilities, bathrooms, or propane fill stations. This was OK, because you parked on concrete, your vehicles parked on concrete, and the grass was just enough for the dogs. We were right on the lake at this spot, and it was cool because we were secluded from the city, but close enough to get out and be able to get and do what we needed to do.

This was also the point when we decided we were going to stay in Texas with the RV, so it was a good time to go and get my car so we both had transportation. I hopped on a one-way flight to Austin, my mom picked me up at the airport, and we met Tim for dinner at one of their favorite places to eat, The Grove. I add this detail because my parents know the owners and most of the staff there at The Grove, and because of this, I know quite a few people that work there too. So, going to dinner here was comforting and felt normal, which I desperately needed. We didn't even need menus because we have the menu pretty much memorized. Normalcy, is it too much to ask for?

We were able to continue making reservations at the RV spot in Flower Mound. However, this RV campground had the rule (because it was beside a body of water) that you could only stay three days. Well, they were not full, so Brandon would check in every couple of days to see if we could extend our

THE PRINCESS AND THE RV

stay. This was a super convenient place for Brandon to be located, because it was literally one of the entrances to one of the mountain bike parks that he and all his buddies would ride their bikes on. You would think I would be irritated at this detail. The location was great for his mountain biking excursions, but what about my needs? Yet at this point, I was so tired of packing up and moving RV parks that I did not even have an opinion on staying at this spot or not staying at this spot. It was on the lake, so I rationalized in my head that at least I could look out and see the water. This went on for three weeks until I finally put my foot down and said I cannot continue to live like this. I needed an address, and I needed some form of stability.

This stability would come in the form of a gated spot with our own storage shed, a doorbell on said gate, an actual address with a mailbox center, laundry facilities, and a fence around our whole space. Why was the fence so important? The dogs. We could let the dogs go to the potty without having to hook them up to the leashes to take them to the potty. The downside of this stability was the location of this gated paradise was located in Princeton, Texas. I remember the first person we talked to about where we were going to park next said, "Oh, Princeton, Texas—meth capital of the world." Great! Perfect!

I am going to just be honest and tell you that I was the only person in that RV park who was high maintenance. What do I mean by high maintenance? By definition: Maintenance—to keep in an existing state (as of repairs, efficiency, or validity): preserve from failure or decline (Merriam-Webster Dictionary Online, s.v. "maintenance," accessed on November 16, 2024 when I was writing this book).

I got my nails done every two weeks; I have already told you my hair appointments were like a "in blood and stone" kind of appointment. I got facials to maintain my youthful glow and liked to drive a nice car and dress in nice clothes with nice shoes. As a side note, gravel driveways, walkways, and walking paths do not do any shoes any favors: I am living proof of this. This all may sound like an entitled person who did not have a clue about living in the real world or someone who cared only for herself. I can tell you that if I did not take care of myself, I would have completely lost my mind. This maintenance gave me a feeling of a normal routine, which I desperately needed in this living situation we found ourselves in.

We moved into Hickman Luxury RV Village on October 6, 2022, and we moved out on July 31, 2023. The term "luxury" was clearly subjective, as our experience did not reflect it.

Before we arrived at the luxury RV village, Brandon needed to take the RV to a dealership to pick up some parts he needed to replace, so that our RV would have a working thermostat and a kitchen sink faucet with enough water pressure to wash our dishes. (We didn't have the luxury of a dishwasher in our RV, so every single dish, fork, cup, and plate had to be washed by hand.) My level of care about things that didn't operate properly in the RV was in the negative caring zone at this point. I can promise you I never in a million years would have dreamed we would live in Princeton, Texas, for this long. God sometimes protects you from the truth, because He knows the truth would undo you. My friend, that truth would have undone me, or maybe it would have given me an end in sight. But regardless, the time we spent at this location created a bond and fierce strength in both Brandon and I that we did not even know we needed.

October meant Brandon's birthday month, and we tried to make it feel special. Our daughter surprised him by coming to stay with us, and we figured out how to turn the back part of our RV into a makeshift bedroom. The two couches we have in the back of the RV can be hoisted up and laid flat to create a bed. I got a mattress topper, sheets, blankets, and pillows to make it feel home-like and semi-comfortable for her.

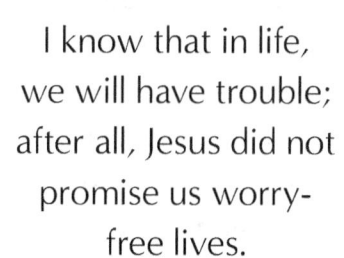

I know that in life, we will have trouble; after all, Jesus did not promise us worry-free lives.

We went to the State Fair of Texas, which is one of Brandon's very favorite

THE PRINCESS AND THE RV

things to do. If you have never been to the State Fair in Texas, you are missing out because it really is bigger in Texas, and you will miss out on things like the half-man show. I don't even know how to describe this event; you must go to believe it and see it. So, you see we did continue to live our lives to escape from our reality, but we always knew where we were going back to.

I know that in life, we will have trouble; after all, Jesus did not promise us worry-free lives. "I have told you all this so that you may have peace in me. Here on earth, you will have many trials and sorrows. But, take heart, because I have overcome the world" (John 16:33). This was one of many Bible verses we kept close to our hearts and would repeat them to each other when we would feel like there was no end to this train we were on.

Days turned into weeks and weeks turned into months. Thanksgiving was approaching, and I was pleading with God to deliver us from the situation. How can this be that we have no home on the holiday for gathering and one that we had historically hosted at our home? God continued to say wait. We had wonderful friends who hosted a Friendsgiving, and we were so blessed by their invitation to join them. These friends were from our old neighborhood who all lived on our old street in McKinney, Texas. It was so much fun to have everyone together again on a holiday and so needed for Brandon and my soul.

All the ladies were interested in how I did laundry and how we cooked in an RV. After living on the same street with these ladies for fifteen years, they knew me well. So, they were genuinely interested in how I did life in an RV. I feel like after explaining the teeny, tiny washing machine situation to them, they immediately understood that the living situation Brandon and I was in was not as glamorous as they initially thought. Their true concern and heartfelt prayers for our situation were deeply touching and very needed during a time of year we all historically were with our families in our home or our families' homes.

Brandon and I also were starting to realize how freeing it was to not have to host the holiday for all the people we usually did. I would say that hosting is part of my DNA, and I truly enjoy doing it, but it is also a ton of work. I never minded because we felt like it set an example for our children. We would invite both sets of Brandon's and my parents, which would include Brandon's mom and stepdad and his dad and stepmom. We would include his mom's sister and husband and their family, and all our siblings and their significant others, which changed many times over the years. We would also invite neighbors we knew who had no close family or neighbors going through a hard time and/or friends who were like family and knew we threw one heck of a party.

It is one thing to go to someone's house and bring your one dish and leave after you have helped clean up. It is a whole other thing to host the holiday at your house and do all the preparations weeks before, and then clean and put away everything afterward. All this to say I would not change all the memories we have from all the Thanksgivings we have hosted over the years; I don't want to forget mentioning my sweet potatoes are pretty famous as well.

However, God was teaching us one experience at a time that we are blessed by others when we give them an opportunity to bless us. We were starting to see that through this pause in life, God was showing us a different way of life that He knew would not have been possible had we been living in a traditional brick-and-mortar home. I wish that during this time we could have known that as painful and hard as this experience was, we were learning things that would help us grow and break free of a life of always doing the same thing during holidays. Breaking free from traditions in your life can be a freeing experience that gives you time to contemplate what truly matters in life.

God give us pauses in life, and it is a choice for us to learn from the pause or continue on the same path as before. If I could give you some advice about life, it would be to recognize the pauses, and ask yourself how you want to move forward in life during this time.

Chapter 15
Oh Christmas Tree

What is the most wonderful time of the year? We all have our own definition of what this means, and we all have a reason we believe our time is the best. For me, it is Christmas. Growing up, Christmas was filled with wonder for me. My parents kept the Santa dream alive as long as possible for my sister and me, and we gathered with friends throughout the season for parties, making all sorts of food and drinks for family and friends. Also, the decorations were the best.

I loved going to our childhood babysitter's house for Christmas, as Ms. Priesmeyer lived in an old, historical home in the town I grew up in and turned her house into a Christmas dream. Now, she did not really do any of the decorating; her daughter, Barbara, was the true elf. Barbara had an antique/Christmas decor gift shop in the back of Ms. Priesmeyer's property. I can't remember if this shop was there when they moved into this house or if they built this shop, but the three-story house would turn into the most beautiful version of Christmas I had ever seen.

There were fully decorated trees everywhere throughout the house and beautiful garlands that hung above the fireplaces and strung along the banister of the staircase. As children, when Rebecca and I would stay with Ms. Priesmeyer (when our parents would go away on vacation) during Christmas time, we would help with decorating and making the fudge and wassail for all the people stopping by to talk about the latest news in the town or those coming to get Christmas gifts from the shop. We would also visit friends in the surrounding towns for dinner parties and gift exchanges for the children. My family and I would bake cookies and sing songs, and I felt like the air during Christmas was just a little bit lighter and more joyful.

The signal that it was almost time for the big day was going to Luling to our friends' mom's house, Amy and Alton, for tamales and chili and attending midnight eve services at our local Episcopal church. I remember believing with my whole heart that I heard reindeer outside my window and Santa on the roof. Christmas was magical.

The first Christmas Brandon and I were together was filled with lots of learning for Brandon about the traditions and magic my eyes held during this season. Brandon did not grow up with Christmas being as magical as mine, so this was an adjustment for him over time.

As Christmas approached us while living in the RV, our therapist suggested I decorate the RV to give me a sense of the season. I could not bring myself to decorate the RV and frankly, the reality that Christmas time was here, and we were no closer to a decision of where we would live propelled me into a sadness I didn't know how to cope with. I knew that I had to change my situation to avoid me becoming a real-life scrooge. So, I packed all my clothes and all my shoes and most of my essentials, loaded the dogs up, and decided to drive south to my parents' house. I packed as if I was not about to come back to that RV any time soon.

My mom tried to cheer me up with the idea of helping her decorate and go to the events they had planned. It was a good attempt, but I was in a place so foreign to me that I literally had to force myself to participate. This should have been a time of true excitement, because Brandon and I would have both kids with us at my parents' house for Christmas. This time should have been a time of being able to count my blessings that my parents had the space to let Brandon, and I stay with them for a while. There were a lot of things that I should have been able to reflect on how blessed we were, but, honestly, I couldn't.

I was so used to the preparation of Christmas that the absence of the holiday preparation left me feeling the absence of Jesus as well. I did not understand why we had no answers, no home, and no direction in which to go. Brandon and I both believe that God wants the very best for us, but that does not mean God is a genie in a bottle. God is our helper and our guide, but we must put in the work to get the answers.

I honestly was praying for a Christmas miracle and truly believed it could happen. But, in retrospect, I was doing exactly what I said God was not. I was treating God like a genie in a bottle and was mad that my bottle was broken. This is so embarrassing to admit, but it is the truth: God does not need our prayers, nor does He need our pity. God wants us to have faith; faith is knowing that God is in control and no matter what the situation is, He will provide and protect.

I now know God was molding and shaping me in this time. I said I had faith, but did I have faith when the cards were stacked against me, and my bottle was broken? Did I have faith when everything that made me comfortable was taken away? Did I show others I had faith despite our circumstances? I talked a big talk, but my heart sang a different tune.

Christmas came and went, and my heart was realizing just how strong it had to be. Sometimes the miracles we experience are the very thing that we want to disappear. The story of Jesus's birth was not a glamorous story of beautiful surroundings and pretty people having parties and opening presents. No, the birth of Jesus is a story of two ordinary people who had faith in what the angel of God told them.

> "Don't be afraid, Mary," the angel told her, "for you have found favor with God! You will conceive and give birth to a son, and you will name him Jesus. He will be very great and will be called the Son of the Most High. The Lord God will give him the throne of his ancestor David. And he will reign

over Israel forever; his Kingdom will never end!" (Luke 1: 30-33, NLT)

The real miracle for the world was born in a manger on straw in the open air. God doesn't call the equipped; He equips the called. Miracles come in all shapes and sizes; our hearts and minds just have to be open to what God wants to show us.

> "It is not that we think we are qualified to do anything on our own. Our qualifications comes from God." (2 Corinthians 3:5, NLT)

Chapter 16
Back to The Basics

Living with your parents growing up is one thing, but living with them as a married couple in your fifties has a completely different meaning. One does not usually enter this arrangement unless necessity insists. My parents live closer to the town I grew up in, so my inner bent of looking like I had it all together went into overdrive as I navigated the questions and people we were surrounded by while there. I went through all the pity party emotions of why me, and then I decided it was time to snap out of it. It was time to find real faith and look around me to see the blessings I was given.

I am a handy girl, and I learned just how handy I was during that time frame of life when we were all forced to pause our life in 2020. I learned how to wire our ethernet plugs for the Internet, and I learned that if YouTube can't answer your question or teach you how to fix a problem, you're better off calling in the professionals. My parents knew this about me, and I became the designated live-in handywoman. A problem would arise almost daily with their house that required a tweak, a fix, a new part, or some old-fashioned elbow grease to eliminate the issue. This was a true learning experience for me about how my parents lived and what they were able to deal with regarding their house.

If a toilet leaked, they called the plumber. Not necessary; I can replace or fix most all parts that would make a toilet leak. Next problem: The sliding glass doors in their bedroom were completely misbehaving, and the door people and I worked together on the phone as I went up and down the attic stairs, where the door communication system was housed, and we fixed the issue. I was pretty much labeled the hero that day.

During this time, my parents also had gotten a precious, little puppy, or so I thought. They had gotten an Australian labradoodle, and he is cute but too smart for his own good. The cute and furry addition to the Von Dohlen household was given the name Coco, because he was cinnamon in color. Coco thinks he is a person and does not understand why he can't just sit at the table with us and eat like normal dogs should. Coco is a true work in progress and with our new living arrangement came with it dog-sitting duties.

My parents travel a lot in general and are very involved in many organizations that take them all over the world. Having a new puppy was a bit of an issue because Coco wasn't up to date on his shots, and the boarding facility my parents used wouldn't accept him until he had received the required vaccinations. So, it was nice that we were living at their house because we could keep their dogs. This was totally fine, because we contributed to the dog population in the house by three, though our dogs were just older and knew the rules of the roost. Coco, not so much.

Coco would decide about 4:00 a.m., that he was done sleeping in the kennel and by done, I mean he would bark and not give up. Most dogs would give up and realize no one was coming to their rescue, but not Coco; he was diligent. He barked until he was let out of the kennel. Coco's kennel was located directly beneath the bedroom we slept in upstairs, so we had a direct line to Coco's voice. This experience will be a forever memory only because of the one time I thought I could get back to sleep by allowing Coco to come upstairs and sleep on the floor right beside me. Joke was on me.

My parents' two dogs are not allowed upstairs, and they know this because they have invisible fence collars on, and a boundary line is located at the base of the stairs. So, if they come too close to the stairs, yep, a beep and a zap. Coco was still little at this time, so I figured it was fine because he had not yet learned the rules. One night when my parents were out of town,

I gave in to Coco's barking. At about 4:20 a.m., I went and got Coco, took off his collar, and brought him upstairs to our bedroom. He was so good and still and, most of all, silent. I fell back into a blissful slumber, but when I awoke, my eyes opened to a white nightmare.

I might have been in a blissful slumber, but Coco had other plans. Coco had silently ripped open the upholstered chair by the bed and pulled out all the stuffing inside the chair, all while I was in my blissful sleep. Like, what dog can do that silently? Coco. I honestly could not even get mad at him, because it was my own fault for bringing him upstairs and thinking he would fall into an angelic slumber. It took me about two months before I told my mom about this incident. So, that just goes to show you that no matter your age, you can instantly be transported to the feeling of being a teenager when you knew you had made a mistake.

Chapter 17
The Memories We Cherish

Blessings come in all forms when you look at life through the lens of God's grace. It was February 1, 2023, in Austin, Texas, and an ice storm hit Texas. We are used to how things respond a bit more north in Dallas, so the precipitation in Austin was not what we were used to, but we had experienced ice storms in our past. This storm, unfortunately, knocked out the power in most of Austin and the surrounding areas.

My parents had a generator, still in the box, but at least they had listened to my plea to purchase one. Brandon and my dad went to get gasoline, and the oil needed to crank up the generator, while my mom and I got out the extension cords and made a game plan of what would be "essential." It was shocking how quickly the temperature dropped inside the house. We all had on layers of clothes and coats, with blankets surrounding us wherever we decided to hunker down for the time.

This was a frightening time because the trees that surrounded my parents' house, and a big percentage in Austin, were big trees, and layers of ice on tree branches do not make for a beautifully happily-ever-after story. The temps stayed in the freezing zone for three days. On one of those days, my parents decided to go to their house a little further down in South Texas to "check" and make sure everything was OK, but I secretly know they just wanted to sleep in the warm house and a warm bed. To say it was cold in their house is an understatement; it was about 45 degrees Fahrenheit inside, with the fireplace in the den at full capacity until the gas ran out.

This was when we found out the natural gas in my parents' neighborhood is shared by the whole neighborhood, and it got filled based on historic usage. Well, this was not a historic event, and every single person was using their gas at full tilt boogie until there was no more boogie in anyone's gas line. The tank in Seven Oaks neighborhood was empty, no more, gone. Guess what the gas company said? "We are swamped; you're on the schedule." The tank got refilled when it was about 50 degrees outside, and the power had finally come back on.

The blessings during this time were that Brandon was there to help my dad fill up the generator tank and knew about when it would run out of gas. Brandon did an excellent job of making sure our generator was working, and anyone else needing gas for their generator got gas delivered to their door via Brandon. While we were waiting on the outside temperatures to warm up, my mom and I cleaned out drawers and closets to pass the time while we had no electricity. This was a fun experience being able to see what qualifies as junk and what qualifies as you just never know.

The worst part of the day was night, because upstairs was so cold and the only bed downstairs was my parents' bedroom, which we did use when they decided to go hide at their warm house and bed down south. We also bought the only remaining space heater in the area at Lowe's. It was technically an outdoor patio-type of heater, but the heating elements were up high, so it posed no risk to our five furry friends.

The ice storm of 2023 was a crazy time and made us all thankful for the safety and shelter of a brick-and-mortar home. The Barbie Dreamhouse was becoming more of a REI handyperson's place to eat, shower, get dressed, and sleep. This storm gave us the appreciation for family who loved us enough to let us live with them with the reality that we were still in limbo, with no idea of where we were going to land. Blessings are what you must focus on when you are going through a storm or valley in your life. The valleys can be lonely and long, so we must continue to look forward, knowing that we won't be in the valley forever. God has a plan, and we must trust Him with it. Easier said than done.

Chapter 18

Lessons in Deception

As the end of February 2023 approached, Brandon had gotten what we thought was a good side income opportunity. It was a complex job that required one to communicate orders of large quantities of chemicals, supplies, or the likes to English-speaking clients. The company was out of South Korea and needed a liaison to take orders and communicate with the client about their needs, then coordinate their orders with the shipping company in South Korea.

At the time, we both were in such a valley, and Brandon and I prayed about this opportunity. It seemed kind of crazy due to the logistics of it, but all the background checked out. With that determined, we decided it was time to get away for a while because Brandon was working on this logistics deal and another deal with a building construction company that wanted to hire him as a consultant to get their operations up and going. We finally had an opening in the sky, and we needed to change our scenery before Brandon got super busy with travel again for work.

We went to Mexico for a week, and it was so glorious. I can tell you when I was packing for our trip, I didn't even care that my wardrobe was ill-equipped for fancy dinners and poolside afternoons. I would figure that out because we were escaping our reality and going to my favorite place on earth, the sunshine and ocean. We arrived in Mexico and our resort to bliss. Brandon and I were in heaven with no cooking and no cleaning, as all our meals were being prepared for us, and there were no appointments to attend: only relaxation was scheduled for us. During this week, Brandon was working on this logistics order with the South Korea company, and it was wild with their communication but going well. The building products company was getting closer to an offer for Brandon, and we really had nothing to come home to, so we decided to extend our vacation for another week. We own a membership with a group of resorts throughout Mexico, Jamaica, and other tropical locations, so we just had to check to see if they had availability at their sister property in Cancun (this particular resort was adults only, so bonus). The resort did indeed have availability, and so we booked the reservation, packed our things, got transportation to our new tropical home for a week, and set off for another week in blissful paradise.

Now, this sounds amazing, right? I mean to anyone who says to you, "Would you like to stay on vacation for another week?" is your response, "No, I must get back home for…" I must be honest and say it threw me for a loop. Remember, I am a planner, and I had not planned on staying another week in Mexico. I didn't pack enough clothes for another week, and I was not sure I had enough of my meds that I take for my kidney/liver disease. I have the hereditary disease my dad had, ADPKD and ADPLD. This disease entails many things, and the main thing I was concerned about at this juncture was my medications and did I have enough. I have worked, with the help of an amazing team of doctors, to be extremely stable with my disease. (I have not talked about that, because that is a whole other ball of yarn.)

Thankfully, I had almost enough meds, and we were in Mexico so the medications I was going to run out of, we could find the equivalent of the medication in Mexico. Mexico does not require a prescription for most common prescriptions, so you can go to most convenient stores to find a pharmacy with a pharmacist on duty to get most any prescription medication.

We moved resorts and prepared for another blissful week of sun and nothing to do. Brandon and I had another amazing, much-needed week of relaxation, but oh, the reality of going home was not a welcome thought to us. However, there was a small light that was starting to shine. We thought we knew what it meant, but we really didn't because we would continue this journey and circle back to the same point.

We saw a house online in the Dallas area that had been on the market since January, and this house seemed to have all the items we prioritized. When Brandon and I are in the market to buy a house, we make a list of pros and cons and rank them as non-negotiable and negotiable items. We have been doing this process since buying our first house, and it helps us to not get emotionally involved in the house-buying process. This would be our ninth house we were going to buy, so we were not beginners at this process.

The house we saw online was having an open house the weekend we would get back from Mexico, so we thought we would drop in after we went to church to see if the online pictures were the reality of the house. We left Mexico with relaxed brains and hearts; it was a good thing, because life was about to throw us some major explosions.

Chapter 19
What Vacation?

We got home from vacation and planned to go to church and then to the open house of this house we found online, as mentioned at the end of the previous chapter. We went to a new church near the house we were going to look at just to see if it was our kind of people. Brandon and I knew that moving to a new area meant we had to find a new church closer to our house where we would be part of the same community we lived in. Unfortunately, this church was not the one, so at least we checked one off the box.

After church, we went to lunch and discussed our list of items by priority before the open house. We wanted to make sure our hearts and minds were on the same page. Then, we finished our lunch, got in our car, and turned off the highway to a smaller road. Brandon and I saw our soon-to-be neighborhood and were immediately in love with the neighborhood and area. We had driven up and down this highway many times, and not once had we seen this neighborhood tucked back into the trees. The neighborhood was not really near any other neighborhoods or in a town per say, but it did have a security gate at the entrance. This was on the list, a silly necessity for me. I was used to Brandon traveling a lot, and the security gate would give me peace of mind being the only person in the house.

This is where I need to tell you that God had this whole plan orchestrated before we could even begin to understand what and why we were going through all of this challenge. I still don't understand all the reasons why we endured what we did, but living in limbo was by design. Remember I told you that when you live in an RV, you have no privacy, and you can't escape each other? We had learned to master the skill of sitting right next to each other but also having the mental separation we required for the introspective moments in life. Brandon and I had learned to be in the same space and still find the quiet moments we craved apart from one another. Living in an RV with your spouse is a set of skills that is learned and not taught. I give you this information because we were learning how to live in our new normal of being together without the chaos of a traveling spouse. We were learning what it meant to wake up every morning together.

Brandon and I walked into this house, and I knew, I knew: This is why we had waited. The house had almost everything on our list of must-haves, and the things it did not have could be added at some point. We had to play it down though, because buying a house is a business transaction. The moment you get emotional is the moment you can really lose control of the business side of buying a house. We knew there were many things that had to happen for this to work out, and if it was meant to be, it would work out.

We toured the whole house and spoke with the listing agent, letting her know we would be in touch. She reached out to us before we reached out to her, telling us that the seller was ready to play ball. Or so we thought. Brandon and I discussed all the things and did make an offer, but the sellers did not even counter, saying we were too far away from their number. Now, we have bought and sold a lot of houses, and never have we dealt with a seller unwilling to negotiate at all. The market had been very hot, but it had drastically changed in the last year, and we were surprised the sellers did not want to counter our offer, or that their realtor had not counseled them to at least show some interest in our offer. This was where the whole "You can't get emotional" thing had to come into play. We knew it was a business transaction, and it would work out if it was supposed to.

The next few days proved that all things do indeed happen for a reason. The job Brandon had with the South Korean company turned out to be a

WHAT VACATION?

scam, and we had lost a lot of money investing in it. I'm talking about game-changer money, and it was just gone. The whole situation was insane, and we were absolutely devastated that this had happened. We are not naive people, but obviously Brandon and I needed to learn another hard lesson in life.

The fact that the house did not work out at this point was a blessing. We were out a significant amount of money and completely lost as to where we should be or where we should go. Were we back to square one? What was God trying to tell us? We tried to remain positive, knowing God had a plan, but we were both drained at this point. The highs and lows of our lives were taking a toll, and we both were just ready for some sort of normalcy. It had almost been one year since we sold our house in Texas, and we were nowhere closer to buying another one than we were a year ago.

Thankfully, we decided to re-group and go to Austin for a while and see what God was telling us. The following Sunday we went to the church we had been attending while we were living with my parents. The name of this church is Austin Ridge and Brandon and I both love Pastor Brad. (If you're looking for a church in the Bee Caves area of Austin, we highly recommend this church). That Sunday we heard a message on finding your spaces in life, and the sermon left us with five questions/commands to ask ourselves to answer: 1) Pray, ask, listen; 2) Where are we located now?; 3) What is our passion?; 4) Where does God want to move you?; 5) Do it with someone else. In other words, when you find your why in life, make sure you are doing it with someone you love.

Those five questions/commands take some serious soul-searching, but my first question was "God, are You telling us we need to live in Austin?" I did not have to wait long to hear what Brandon thought. We got in his truck after church, looked at each other, and both knew we were thinking the same thing. "Did you feel like God was telling you that we should live in Austin?" I could barely answer because I was crying and shaking my head yes.

This was crazy though. We had been so focused on finding something in Dallas that the thought of staying in Austin had not even crossed our minds. Were we crazy and just trying to make something out of nothing? No, we both said that if we did not act on a calling from God that we both had distinctly heard, then we would regret the decision for many years to come, not to mention miss out on whatever it was God was calling us to do. We talked about all the things involved in living in Austin and what it would look like, etc.

Then we connected with a realtor and began to look at rental property, because we had no idea where we wanted to live. We looked in Bee Caves, Lakeway, Spicewood Springs, Dripping Springs, and beyond, finally finding a house in Spicewood Springs that we thought would work. We put the deposit down and went back for a final tour before signing the paperwork. And when I say I had a feeling of discontent, turmoil, and fear that overwhelmed me when we were in that house, I'm not exaggerating; it was intense. I can't really explain what it was except that I knew that this was not the place for us. It did not make sense. We had discussed all the options, and we thought it made sense on paper, but the feeling of darkness overwhelmed me while I was inside that house. So, that was not the one. We continued to look, but nothing was clicking. Our realtor was working hard to find us a place that worked within our specifications.

Had we missed the message? We both heard God speaking the same thing to us that day. What in the world was going on? Why were we going in this circle of confusion and disarray? To not know where you are supposed to be in life is a very lonely feeling. It really makes you dig deep and be honest with yourself about who you are, what you want in life, and whether you are proactively doing something to achieve the answers to these questions. We truly thought we knew where we were supposed to be going, but apparently God had other plans, so we had to continue to be patient.

After a few weeks of Brandon going back and forth between Dallas and Austin, we needed to discuss what we wanted to do, because this lifestyle of us both driving back and forth was truly exhausting and frustrating. Brandon was continuing to network in Dallas, and the connections he was making

WHAT VACATION?

in Dallas seemed to be more lucrative than anything that was connecting in Austin. However, we were very gun shy to make any rash decisions based on our previous thought process about moving to Austin. The only thing we knew for sure was that God had us on a journey, and we were learning a lot about each other, a lot about ourselves, and the conclusion of both things led us to believe that He was in control, and we were not. We continued to pray for a sign, for direction, and for ears to hear and eyes to see. It seemed like it was much longer than six weeks of thinking we were going to live in Austin, but it was only a period of six weeks. This is proof that whatever you are going through in your life may seem insurmountable and never-ending, but there is always an end in sight. No season lasts forever, and no storm continues to rage on in our lives. We do come out of the storm, and we begin to pick up the pieces to create beauty from the broken.

Chapter 20
Where In the World Are We?

At this point, I think our children thought we had lost our minds, and so did many of our friends. Some of our family even doubted our decisions, and the questions they asked began to make us second-guess our decision to bring it back to where we started. The only author of our story is our God, and we had to trust this more than any voice we heard around us. After all, we were the ones living on this roller coaster, and we had to make the decisions based on our reality, not anyone else's thoughts, feelings, or opinions.

It was hard to tell my mom that we were going to focus more on Dallas than Austin, and that we both felt like God was guiding our paths. Moms have a lot of wisdom, and my mom is no different. She knew the reality of us living in Austin was too good to be true, but the moment was fun to think about while it lasted. No matter what we decided, my parents just wanted us to be happy and for us to find a place where we could plant our roots and grow.

The month of May 2023 was filled with many trips from Austin to Dallas and Dallas to Austin. If you have ever had the unfortunate experience of driving on I-35 between Austin and Dallas, you know that the experience is nothing short of stressful and frustrating. Brandon and I tried to plan our trip so that we left either city early in the morning and were on the interstate before the work traffic began to get crazy. When you are living in a life situation that has you in a perpetual state of limbo, you learn the skill of adaptability, or you spiral into an abyss of the unknown. We chose adaptability and tried to make each day better than the last.

Remember that house I told you guys about, where I said I just knew it was for us (the house that we made an offer on, but they said no)? Well, it just so happened that it was still on the market, and the current owners had lowered the price four times by this point. They were having open houses every weekend, which signaled to us that they were very motivated to sell at this point. This is a bonus for buyers when the sellers are motivated.

I was continuing to supervise the listing from afar, knowing it felt right but realizing that the details were not in my control. It was a Saturday afternoon, and Brandon and I were back in Austin. We were in the pool at my parents' house, and I said, "Why don't we make another offer on the house in Crossroads?" Brandon was aware of all the statistics about the house and the fact that it was still on the market. He asked me if I was sure I still liked it. Did we need to go see it again? Many things had transpired since we last saw this house. My response was an instant, "I knew from the moment we walked into that house that it felt like ours." Question answered.

On June 15, 2023, we made another offer on the house in Crossroads. Do you see what I am doing here? I was trying to keep it business and avoid getting personal, so that my heart did not get involved. We had a number in mind that we were willing to go to, but that was a firm number, and if it did not work out this time, then it was not meant to be. The prayers that we were praying were intense and specific. "Lord, if this is the house You mean for us, make it so. Lord, if this is not the house for us, then take it away and show us our next steps." God is always listening, and He always has an answer; it just might not be the answer we want to hear.

After some back and forth of negotiating, we agreed on a purchase price on this house, and we were set to close in two and a half weeks from June 17, 2023. Yes, fast, but I had done this before, and oh was I ready and prepared for this. Buying a house can truly be overwhelming no matter how many times you have done it. The next days were spent with appraisers and inspectors and approvals and reports. Brandon and I both knew that all those things can make or break a sale, so we had both eyes open, and each day was taken in stride. Once all the paperwork was complete, and all of the terms had been addressed, we were set to close on July 5, 2023.

There is something to not forget or dismiss when you are going through a trial in life. Don't miss the wait. Don't waste the wait. The wait taught me that I could do things I never dreamed I could do, like living in a tiny space with my husband and three dogs. The wait taught me that I could live without all the things I thought I needed to survive. The wait taught me that I am certainly not in control, and until I realized this, the time was wasted: wasted on worry, despair, frustration, and sadness. I learned that life throws us curveballs just when we think we have it all figured out.

However, I hope that we taught our children that strength comes from perseverance and working together, no matter the situation. I hope we taught our children what a strong marriage looks like. And I hope we taught those around us that looks can be deceiving, and strength and grit are not always visible with the naked eye.

I will never forget having to wash clothes in the little, bitty washing machine and then having to empty the gray tank so that water did not overflow and start leaking out of the bottom of the RV. I will never forget cooking in the kitchen of that RV, thinking how to do all the things without a garbage disposal or dishwasher or much space. But you know what? I did it; we did it. We lived in an RV for seventeen months of our lives, and we didn't die. We didn't die but survived. We both have battle wounds, both physically and mentally, from this experience, and that is OK. We should never go through times of strife in our lives with our eyes closed and our brains locked.

We need to embrace every moment in our lives as opportunities to grow, learn, and realize that we can do hard things, even when we don't plan on doing them. We can do hard things even when we do plan on them, and then those things turn into complete and total chaos.

Life is not one big war; it is many tiny battles fought with newly acquired weapons after each one. The victory does not lie in the win; the victory lies in the decision to get up and put on the armor of God so we can come out of the battle stronger and more resilient humans. "Put on the whole armor of God that you may be able to stand against the wiles of the devil" (Ephesians 6:11, NKJV).

Chapter 21
Being Honest

The question I get asked all the time is, "Would you do it again?" My answer is an immediate yes. Now, I am not saying I want to relive the experience, and I will qualify that yes with a resounding "Be careful what you sign up for." I know that seems insane, because the experience was not always filled with rainbows and sunshine, but the lessons I learned and the growth I had would have been impossible for me in something that was less drastic.

God knew He had to completely take me out of my comfort zone to get me to listen. My comfort zone is a hundred percent inside our home with the comforts that make our lives ours. God had to completely shake up our lives to have us open our eyes to the path He wanted us to be on. As I type these words, I can honestly tell you we still don't know the complete picture of what God is wanting us to do, but we do know we are in a much better space to listen to Him.

Sometimes we must accept the situation and take the focus off ourselves and focus on others to be able to truly find meaning in the wait. Focusing on others can be hard when you're in a life gap that has you uprooted from all your comforts and knee-deep in the unknown discomforts of the world. I had to force myself to continue to be social and part of my people who know me. I had to force myself to get out of my hiding hole and live life even though it looked different than anything I had ever experienced. There are a handful of people who really knew what we were dealing with, and those people are

THE PRINCESS AND THE RV

our people. They met us right where we were. They loved us; they prayed for us; they called us; they texted us; and they made a point to see us. If you have people like that in your life, don't lose them. Hold on and love them and do life with them and make memories. And no matter what life throws at you, love your people.

We live in a world that is flooded with social media feeds and what everyone is doing or not doing. If we are not careful, we can look at social media and go down a very dark road. Their lives seem so perfect: they have the perfect spouse; their children are beautiful; they have the house, the cars; and they take unbelievably amazing vacations. Y'all, that is a filter. Most people show you what they want you to see. They don't show you their bedrooms filled with unfolded clothes, dirty clothes piled on the treadmill, and a freezer filled with enough chicken nuggets to feed a small village.

Life is hard, and it really does not matter what age group you land in, because each group has its own set of difficulties. We all have baggage that is a result of our upbringing, our family and culture, our jobs, our experiences, our sick relatives, our own health issues, and so much more. The challenge is to pull yourself away from social media and just live life. Live life and don't be so focused on other people's lives that you forget to live your life (this includes people you know and those you don't and probably never will). Life can be overwhelming at times, and trying to live other people's lives will keep you in a disillusioned reality. You will miss the beauty right in front of your eyes.

So, get up in the morning, even before you must, make yourself some coffee, and be still with the day. I begin the day with my Bible, my various studies I may be doing in my Bible app, and my caffeine (coffee, an energy drink, or diet soda—no judging, remember). I value this time and look forward to the moments I get to spend with Jesus and my mind, listening to what the Holy Spirit is saying to me. This first thing in the day sets the tone for my day, and I can tell you with complete honesty that if I do not get this time, I am not the best version of myself.

I encourage you to carve out time first thing in the morning to align yourself and your day, as I promise you will not regret it. I also know that we all lead incredibly busy lives. We have families, kids, spouses, houses, appointments of all kinds, jobs, and someone still must go to the grocery store so that our families can eat. I get it; it is a lot, and some days it can be more than one person can handle. Be kind to yourself. Allow yourself the freedom to have some time for yourself to set the tone of your day.

And while I'm talking about time for yourself, also carve out time for self-care. Self-care looks different for everyone, but you know what it looks like for you, so do it. Self-care is so important, and when this is not addressed, our loved ones are the first to notice and take the brunt of our neglect.

I tell you all these things because even though we were living in an RV and we had so many unknowns in our lives, I still had my time in the morning and my self-care carved into my days. Now, I will tell you that it took a minute to be able to figure out how to have my quiet time in an RV with our dogs and my husband. Thankfully, I figured out quickly that if this was going to happen, I had to get out of bed before I had to. Before my husband woke up and before the dogs were allowed to act like their world was ending because I had not let them outside and fed them yet. (Don't worry; they always had water, so they were fine.) I developed a newfound love for watching the sunrise. I have always loved sunsets and sunrises, but don't often get the opportunity to see the sun rise because I am usually in a room far from a window early in the morning. The couch in our RV is right, and I mean right, by the window that faces east; (This was when our RV was parked in Princeton, which is where we stayed in our RV the longest) so I had an awesome vantage point. I saw some of the most beautiful sunrises while living in our RV. God's gifts are all around us when we choose to open our eyes to them.

It truly is amazing the little things you can learn to appreciate if you allow yourself the time to slow down, watch, and feel the beauty right outside your door. That couch area was my quiet time area, our kitchen, my yoga room,

THE PRINCESS AND THE RV

our TV-watching hangout, my office, where we had Zoom calls, and where we ate dinner. We had learned how to live in small spaces. Now, I did not say we learned to love living in a small space; I just said we learned.

Our traveling home taught us many lessons along the road. We learned that you really need to have some form of GPS device that knows your length and height, so you don't crash into any bridges. Walkie talkies are essential for communicating with each other when pulling into any space with your RV. Yelling out directions and letting each other know just how far away you are from the tree, or the neighboring RV is not good RV etiquette. Nobody wants to hear this, we promise.

Not all laundry facilities are equal and that driving past the point of being tired is a bad idea. You should get something to eat before you set up your RV space; just trust us on this one. Just because you are not an RV person, people are so willing to help you learn to be one. Also, YouTube is your best friend when owning an RV and that just when you think you have it all figured out, you don't. Brandon and I learned that without each other, we would have never survived this life experience. Starting marriage counseling while living in an RV was one of the best decisions we ever made. So, take the trip, enjoy the ride, and don't be scared to try new things. Just know that it might not work out like you think it will, and that is OK.

Chapter 22
A Permanent Address

The day of July 5, 2023, was the day we closed on our new house. I had painters, electricians, plumbers, and carpenters ready to roll on July 6, 2023, and after one heck of a crazy week, we were about ready for movers to arrive at our new house. I do say house because it takes a while for a house to feel like a home. It took five eighteen-wheelers and ten men to help Brandon, and I begin to add space to our lives. It took three days to unload our stuff and three days for Brandon, myself, and four of the moving crew guys to unpack our boxes. Now, I am very aware that having people help unpack your boxes is not the norm for everyone. I will just say that there are some things we are willing to pay for in life.

I also met the sweetest, little angel during this unpacking process. Letty was a gift from above, a skilled unpacker who figured me out quickly. I may look like I don't like to sweat or break a nail but looks can be very deceiving because I can work and keep up with the best of them. Letty and I worked tirelessly day in and day out, as she was literally my right-hand girl during this process. She had a grandmother who instilled in her the importance of making a bed correctly the first time. I love her grandmother for this.

Letty helped me wash and dry sheets, pillowcases and blankets, and together we made every single bed in our house. Side note, a well-made bed with all the pillows and throw blankets to match is the center piece of any bedroom, and I take this look, feel, and comfort of a bed very seriously (as is

THE PRINCESS AND THE RV

evident by all the pillows one must take off the bed, in our home, before you can even pull down the covers to get into the bed). There is just something about having a bed made that helps me not feel overwhelmed, especially in the disarray surrounding me when I am unpacking and organizing.

Letty had been a blessing for me during this time, but not all things can be put away without the knowledge of how the people do life. Once Letty did all she could do for us, she and I broke down the remaining boxes, and she loaded them up in her moving van and away she went.

As all the items were now out of the boxes, it was time to start putting things away in our new house. I became like a robot and worked from the moment the sun rose until it was dark. Some nights I was so tired, the thought of taking a shower was overwhelming. I consoled myself by jumping in the pool and having the pool chemicals be my soap. Each day felt like the last, and I felt like the progress I was making was non-existent. Organization is one of my gifts and for this, I am thankful.

I am also able to see the value in things or the lack of value. We had so many black trash bags filled that it became comical on trash day with the amount of trash we had; this went on for a good two months. We had trash bags filled with actual trash, trinkets that no longer felt important, and duplicate items that had lived out their usefulness. We both felt like we were settled enough at this point to slow down and enjoy our space. Brandon and I both were like a fish out of water, because we had not had a home in so long, and it did not yet feel like our home yet. We had work still to be done on the house, but the most exciting part of making your house a home is in the memories you create and the dreams you know that are still to come. We had a house, a real place to put our heads down with real walls, and a roof that did not threaten to implode when a big Texas rainstorm pounded down from the sky.

You may be wondering where that RV was after we moved into the house. Well, not to worry; we had a plan. I sure hope by now you have realized the importance of the best-laid plans need to come with a whole lot of humility and flexibility. We had until the end of July 2023 to pack up our spot and pull the RV out of the luxurious town of Princeton, Texas. Brandon really did most of this work, as he knew how done I was with all things RV. He secretly was probably scared that I would throw everything away. Whatever the case, I stayed at our new house and worked while Brandon packed up the RV to bring it to the house for us to unload what we needed out of it. We needed to unload the refrigerator and freezer for sure, and we needed to make sure that all the other items were secured or properly stored before taking the RV to live at the dealership to have the service items addressed.

We got this done, and I waved goodbye to the RV as Brandon pulled it away from our driveway on August 1, 2023, taking it to be dropped off at the dealership for that long list of repairs. Most of these repairs required a lift to get underneath the RV and then something to get on top of the RV to fix these repairs. If you ever must deal with an RV service place, just know that they will promise you the sun, moon, and stars but will absolutely fail to deliver (as you learned at the start of our RV adventure). I just say this to warn you that if you're on a timeline, be prepared to have some wiggle room in your timeline for picking up the traveling little home on wheels.

As I type these words almost one year later, guess where that RV is? Not at some awesome spot in Colorado waiting for us to visit. No, it is still at the dealership, but that is another story waiting to be written.

Epilogue

Brandon and I are settled in our new home and have had almost a full year of holidays to celebrate with family and friends. We enjoyed having our daughter live with us for about a month last fall and winter. This came as a surprise for us all due to a health issue our daughter had that required her to be at home for us to monitor her. However, this was a special time to have, as we knew this would not last forever, and we also knew that the timing was also a blessing. Her being with us would not have been nearly as amazing had we not been living in a house during this time. We have hosted many dinners and even had all our daughters' friends from high school spend the night a few times. We hosted Thanksgiving and about forty people joined us for lunch, dinner, and then the after-party to watch football. This was the Thanksgiving that Dolly Parton was at the halftime show, and I was thoroughly impressed with all things Dolly and all things Dallas Cowboy cheerleaders.

Once the kids headed back to school, it was officially time to decorate for Christmas. It had been two years since I had decorated, and I was so ready to bring our house to life. We thoroughly enjoyed having Christmas and all our family under one roof. I really went all out in decorating for Christmas, and I enjoyed every single second of it, marking that Christmas as one of the best I can remember in a very long time. Something about being out of your comfort zone during the holidays brings it full circle when you get back to the roots of comfort.

Brandon and I are both figuring out what life is like being together every night. I am learning how to cook almost every night of the week again, and I am OK with that; I have a dishwasher and a garbage disposal again. We both have an office to have our quiet time now, but every now and then, we both glance at each other and miss the closeness we had in the RV in spirit. I don't miss the space; I miss the moments we had that will forever leave an imprint on my heart. We are both independent business owners, and I am just now getting all my items ready to re-launch my business again.

Life is full of so many twists and turns, and if we don't have a solid foundation to stand on, we will get swept up. I encourage you all to embrace the unknown, try new things, and go on the adventure. I promise you will not regret the decision, and you will bloom in ways you never knew you could. Your new self will thank you for taking the unexpected adventure, and you might just find independence from weakness in your life.

To all my readers: Thank you! No, I mean it. Thank you so much for taking your precious time to read about an adventure of two people who thought they had it all figured out. You are why I wrote this book. I knew I had to tell the world about how when life throws you curveballs, you must be prepared to catch them or go chase after them if you miss. I hope that you laughed while you read this book, that you got insight into who I am as a person, and, finally, I hope this book gives you the courage to step outside your comfort zone and try new things.

I can promise you that you will stumble and might even fall in life, but you will get back up and won't be able to go back to the old you. I'm not so sure I am the one who will talk you into buying an RV while not having a house to go back to, but I will be the one cheering you on with whatever you decide. You can count on me to be honest about the realities of living in an RV on an unplanned adventure that helped me overcome my weakness of trying to live in perfection. Spoiler alert: Perfection is not possible in life, and isn't that the beauty of life? I also pray that when life has you living through an unexpected situation, (because it will at some point) you reach out to your people. Please do not go through life challenges alone, God puts people in our lives to help us navigate the peaks and the valleys. Seek out your people, they are the blessings that make your life so beautiful.

To my husband: You are my everything, and I say this with every fiber of who I am. We have grown so much through the years, and I am so thankful I have you by my side as we continue to see where life takes us. Thank you for being my biggest fan while I was writing this book. Thank you for understanding when I had to revisit moments during our RV experience that were hard to revisit, and I carried the mood and feelings with me. I truly don't know if this book would have gotten started had it not been for your gentle nudges and bragging to anyone, we met that I was writing a book.

Thank you for allowing me to be a hundred percent authentic and myself. I love you and am honored to be your wife. I can't wait to write about the next pages that our life writes—just no more RV adventures.

To our children: You both are truly amazing people, and we are humbled to be your parents. We know it was hard to see your parents try to navigate life with no certainty and no plan to fall back on. I hope that you learned from our journey, and I hope that you saw marriage is so much more than the words you say on your wedding day. Colton, you now have Lindsay to include in each decision you make, and I pray that you both will remember to dig deep when times get hard. Logan Elisabeth, you are soon to be married to Sloan, and I pray that you both take each day to remember how very blessed you are to have met the other half of your heart. Marriage is so much more than the wedding and the vows, and I pray all four of you can draw strength from our marriage. Your dad and I were tested beyond what we signed up for, but we did it. We maybe did not do it eloquently, but we did it. I hope the memories you both have of the RV will be reminders to you for the difficult times in your life that will surely come. I pray that above everything in your life, you put God first because without Him, life does not work. I love you both more than I can express in words. Life is waiting for you both; go and live it with purpose and passion. It is worth the sacrifice.

To my parents: Thank you! I know that navigating this journey with us was not easy. I know that as a parent, you feel obligated to have the words and the solutions to life's questions. You both helped us in ways that we did not even know we needed, and you both made this journey bearable in the times that we thought we could not go on one more day in the unknown. Thank you for always exhibiting unconditional love, and for always setting the bar for excellence in the most genuine and loving way. I love you both and am so thankful God chose you both to be my parents.

To all our friends: You are the best! We literally could not have done many days without your lifeline for us. You sheltered us; you loved us; you fed us; you did not judge us; and you let us navigate this journey with the promise that you were always there when we fell and needed to be lifted back up. Thank you. Your prayers and support and phone calls and texts were invaluable to me. I am blessed to have people in my life who know me and know when I am not okay; and for that, I am eternally thankful to you all.

To my editor: Your ability to help me find the missing components needed for a reader to truly understand my point is amazing. Blair, you were encouraging and real and understanding through this whole process. You helped me take my story from pretty good to something I am truly proud to put my name on. I could not have dug as deep into the details of my story without your gentle nudging and asking me to clarify many situations in my life and on our RV adventure. You gave me the confidence to be one hundred percent authentic. I knew what happened, but you helped me understand the importance of my reader understanding each and every situation. You helped my story bloom into something really beautiful, and I could not have ever done this without your help and true empathy. You will definitely be getting a call when the next book takes shape. From the bottom of my heart, thank you.

At the top of the UTV adventure in Ouray, Colorado

Tripp, Kona and Savannah Grace in their car seats on our adventure

There she is, all 40 feet of her in upstate New York

Brandon and I at Sun Palace in Cancun, Mexico

This is right before going to the Secret Stash in Crested Butte, Colorado to learn about the offer from the company in Pennsylvania

A moment we took while stopped in Pittsburgh, Pennsylvania

Savannah Grace serving as the CanAm mascot

The bar at the Ritz-Carlton in Philly

A deceptive smile while I was packing our RV outside our house in McKinney, Texas

The mattress that served as protection for the glass doors from the CanAm, and eventually got gifted to an RV park in Ohio.

The glorious laundry room with the big machines

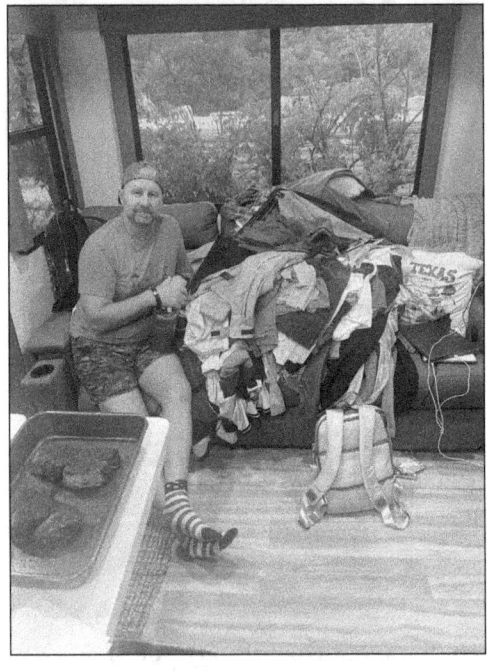

Brandon wondering where all our (mostly mine) clothes are going to go.

My first breakfast I cooked on the blackstone

The turkeys from the RV park in Little Elm, Texas

A double rainbow on our adventure

The Sherpas for the Leadville 100. Pictured are:
Claire Cowgill, Tanya Cowgill, and myself

Brandon and I after the Leadville 100 mountain bike race in Leadville, CO

Our RV and truck parked in Ouray, CO

"The Princess and The RV: Finding Peace in Life's Detours"
Spotify Playlist

1. Hayden Calnin, Introduction; Nothingness, Cut Love PT. 1, Hayden Richard Calnin, Hayden Calnin, Network Music Group, Gage Music, 2016, Spotify

2. Your World Within, Stronger Than You Think You Are, Resilient, Your World Within, Eddie Pinero, Your World Within, 2016 Spotify

3. Ashley Cooke, It's Been a Year, Shot in the Dark, Will Weatherly, Ashley Cooke, Brett Tyler, Jimmy Robbins, Big Lous, Warner Chappel Music, 2023, Spotify

4. Chris Lane, Lauren Alaina, Dancin' In the Moonlight (feat. Lauren Alaina), Single, Brett Tyler, Jesse Frasure, Sherman Kelly, Joey Moi, Big Loud, 2022, Spotify

5. Tim McGraw, Damn Sure Do, Here on Earth, James T. Slater, Tony Lane, Byron Gallimore, Tim McGraw, Big Machine Records, LLC, 2020, Spotify

6. Priscilla Block, Hillary Lindsey, I Know A Girl, Welcome To The Block Party, David Garcia, Hillary Lindsey, Priscilla Block, UMGN InDent Records, Concord Music Publishing, Spirit Music Group, 2022, Spotify

7. Andy Grammer, Good To Be Alive (Hallelujah), Magazines or Novels, Ian Kirkpatrick, Ross Golan, Andy Grammer, Ryan Met, S-Curve Records, Warner Chappell Music, 2014, Spotify

8. Ben Rector, Old Friends, Magic, Ben Rector, John Fields, OK Kid Recordings, LLC, 2018, Spotify

9. Ed Sheeran, Castle on the Hill, ÷ (Deluxe), Benjamin Levin, Ed Sheeran, Benny Blanco, Ed Sheeran, Atlantic Records UK, Hipgnosis Songs Group, Kobalt Music Publishing, 2017, Spotify

10. Mercy Me, Even If, Lifer, Ben Glover, David Garcia, Bart Millard, Crystal Lewis, Tim Timmons, Ben Glover, David Garcia, Fair Trade/Columbia, Capital CMG Publishing, Spirit Music Group, 2017, Spotify

11. FINMAR, Homegrown, Homegrown, Alex Marshall, Cameron Bedell, Thomas Finchum, Alex Marshall, Thomas Finchum, FINMAR, 2021, Spotify

12. Need To Breathe, Testify, Hard Love, Bear Rinehart, Bo Rinehart, Ed Cash, Need To Breathe, Atlantic Records, 2016, Spotify

13. Onoleigh, Legacy, Legacy, Oneleigh, Armen Paul Arakelian, Emile Ghantous, Josh Goode, Onoleigh Pommier, Sam SZND, David Dorn, Onoleigh, 2025, Spotify

14. Tasha Layton, Look What You've Done, How Far, AJ Pruis, Keith Everette Smith, Matthew West, Tasha Layton, Keith Everette Smith, BED Recordings, 2022, Spotify

15. Alex Warren, Ordinary, You'll Be Alright, Kid (Chapter 1), Adam Yaron, Atlantic Records, Hipgnosis Songs Group, Warner Chappell Music, 2024, Spotify

16. Teddy Swim, You're Still The One, You're Still The One, Robert John "Mutt" Lange, Shania Twain, David Cobb, Lee Rouse, Warner Records, 2020, Spotify

17. Carrie Underwood, Favorite Time Of Year, My Gift (Special Edition), Carrie Underwood & David Campbell, Chris DeStefano, Hillary Lindsey, Carrie Underwood, Greg Wells, Capital Nashville, Concord Music Publishing, Sony Music Publishing

18. Carrie Underwood, Mary, Did You Know?, My Gift, Carrie Underwood & David Campbell, Buddy Greene, Mark Lowrey, Greg Wells, Capital Nashville, 2020, Spotify

www.ingramcontent.com/pod-product-compliance
Lightning Source LLC
LaVergne TN
LVHW021225310825
819866LV00009B/109